The Ultimate *Quotable*
EINSTEIN

The Ultimate Quotable EINSTEIN

COLLECTED AND EDITED BY
Alice Calaprice

WITH A FOREWORD BY
Freeman Dyson

PRINCETON UNIVERSITY PRESS

PRINCETON AND OXFORD

Published by Princeton University Press, 41 William Street,
Princeton, New Jersey 08540

In the United Kingdom: Princeton University Press, 6 Oxford
Street, Woodstock, Oxfordshire OX20 1TW

press.princeton.edu

Library of Congress Cataloging-in-Publication Data
Paperback ISBN 9780691160146

Einstein, Albert, 1879–1955.
The ultimate quotable Einstein / collected and edited by Alice
Calaprice ; with a foreword by Freeman Dyson.
p.   cm.
Includes bibliographical references and index.
ISBN 978-0-691-13817-6 (hardcover : alk. paper)   1. Einstein,
Albert, 1879–1955—Quotations.   I. Calaprice, Alice.   II. Title.
QC16.E5A25 2010
530.092—dc22   2010002855

British Library Cataloging-in-Publication Data is available

This book has been composed in Palatino

Printed on acid-free paper. ∞

Printed in the United States of America

3 5 7 9 10 8 6 4 2

For my growing extended family—

The Abeghians, Braunsfurths, Calaprices,
Hazarabedians, Whittys, and Wongs,

especially my sweet grandchildren,

Emilia and Anya Calaprice, and
Christopher and Ryan Whitty

Contents

Foreword

My excuse for writing this foreword is that I have
been for thirty years a friend and adviser to Princeton University Press, helping to smooth the way for
the huge and difficult project of publishing the Einstein Papers, a project in which Alice Calaprice has
played a central role. After long delays and bitter
controversies, the publication project is now going
full steam ahead, producing a steady stream of volumes packed with scientific and historical treasures.

I knew Einstein only at second hand through his
secretary and keeper of the archives, Helen Dukas.
Helen was a warm and generous friend to grown-
ups and children alike. She was for many years our
children's favorite babysitter. She loved to tell stories about Einstein, always emphasizing his sense of
humor and his serene detachment from the passions
that agitate lesser mortals. Our children remember
her as a gentle and good-humored old lady with a
German accent. But she was also tough. She fought
like a tiger to keep out people who tried to intrude
upon Einstein's privacy while he was alive, and she
fought like a tiger to preserve the privacy of his
more intimate papers after he died. She and Otto
Nathan were the executors of Einstein's will, and
they stood ready with lawsuits to punish anyone
who tried to publish Einstein documents without

their approval. Underneath Helen's serene surface we could occasionally sense the hidden tensions. She would sometimes mutter darkly about unnamed people who were making her life miserable.

Einstein's will directed that the archives containing his papers should remain under the administration of Otto Nathan and Helen until they determined it was time to make a transfer, and should thereafter belong permanently to the Hebrew University in Jerusalem. For twenty-six years after Einstein's death in 1955, the archive was housed in a long row of filing cabinets at the Institute for Advanced Study in Princeton. Helen worked every day at the archive, carrying on an enormous correspondence and discovering thousands of new documents to add to the collection.

In early December 1981, Otto Nathan and Helen were both in apparently good health. Then, one night, when most of the Institute members were on winter holiday, there was a sudden move. It was a dark and rainy night. A large truck stood in front of the Institute with a squad of well-armed security guards in place. I happened to be passing by and waited to see what would happen. I was the only visible spectator, but I have little doubt that Helen was also present, probably supervising the operation from her window on the top floor of the Institute. In quick succession, a number of big wooden crates were brought down in the elevator from the top floor, carried out of the building through the open front door, and loaded onto the truck. The guards jumped on board and the truck drove away into the night. Before long, the archive was in its final resting place in Jerusalem. Helen continued to come to work at the Institute, taking care of her correspondence and tidying up the empty space where the archives had been. About two months later, suddenly and unexpectedly, she died. We never knew whether she had had a premonition of her death; in any case, she made sure that her beloved archive would be in safe hands before her departure.

After the Hebrew University took responsibility for the archive and after Otto Nathan's death in January 1987, the ghosts that had been haunting Helen quickly emerged into daylight. Robert Schulmann, a historian of science who had joined the Einstein Papers Project a few years earlier, received a tip from Switzerland that a secret cache of love letters, written around the turn of the century by Einstein and his first wife, Mileva Marić, might still exist. He began to suspect that the cache might be part of Mileva's literary estate, brought to California by her daughter-in-law Frieda, the first wife of Einstein's older son, Hans Albert, after Mileva's death in Switzerland in 1948. Though Schulmann had received repeated assurances that the only extant letters were those dating from after Mileva's separation from Einstein in 1914, he was not convinced. He met in 1986 with Einstein's granddaughter, Evelyn, in Berkeley. Together they discovered a critical clue. Tucked away in an unpublished manuscript that

Frieda had prepared about Mileva, but not part of the text, were notes referring with great immediacy to fifty-four love letters. The conclusion was obvious: these letters must be part of the group of more than four hundred in the hands of the Einstein Family Correspondence Trust, the legal entity representing Mileva's California heirs. Because Otto Nathan and Helen Dukas had earlier blocked publication of Frieda's biography, the Family Trust had denied them access to the correspondence and they had no direct knowledge of its contents. The discovery of Frieda's notes and the transfer of the literary estate to the Hebrew University afforded a new opportunity to pursue publication of the correspondence.

In spring 1986, John Stachel, at the time the editor responsible for the publication of the archive, and Reuven Yaron, of the Hebrew University, broke the logjam by negotiating a settlement with the Family Trust. Their aim was to have photocopies of the correspondence deposited with the publication project and with the Hebrew University. The crucial meeting took place in California, where Thomas Einstein, the physicist's oldest great-grandson and a trustee of the Family Trust, lives. The negotiators were disarmed when the young man arrived in tennis shorts, and a friendly settlement was quickly reached. As a result, the intimate letters became public. The letters to Mileva revealed Einstein as he really was, a man not immune from normal human passions and weaknesses. The letters are masterpieces of pungent prose, telling the sad old story of a failed marriage, beginning with tender and playful love, ending with harsh and cold withdrawal.

During the years when Helen ruled over the archive, she kept by her side a wooden box which she called her "Zettelkästchen"—her little box of snippets. Whenever in her daily work she came across an Einstein quote that she found striking or charming, she typed her own copy of it and put it in the box. When I visited her in her office, she would always show me the latest additions to the box. The contents of the box became the core of the book *Albert Einstein, the Human Side*, an anthology of Einstein quotes which she co-edited with Banesh Hoffmann and published in 1979. *The Human Side* depicts the Einstein that Helen wanted the world to see, the Einstein of legend, the friend of schoolchildren and impoverished students, the gently ironic philosopher, the Einstein without violent feelings and tragic mistakes. It is interesting to contrast the Einstein portrayed by Helen in *The Human Side* with the Einstein portrayed by Alice Calaprice in this book. Alice has chosen her quotes impartially from the old and the new documents. She does not emphasize the darker side of Einstein's personality, and she does not conceal it. In the brief section "On His Family," for example, the darker side is clearly revealed.

In writing a foreword to this collection, I am forced to confront the question whether I am committing an act of betrayal. It is clear that Helen would have vehemently opposed the publication of the intimate letters to Mileva and to Einstein's second wife, Elsa. She would probably have felt betrayed if she had seen my name attached to a book that contained many quotes from the letters that she abhorred. I was one of her close and trusted friends, and it is not easy for me to go against her express wishes. If I am betraying her, I do not do so light-heartedly. In the end, I salve my conscience with the thought that, in spite of her many virtues, she was profoundly wrong in trying to hide the true Einstein from the world. While she was alive, I never pretended to agree with her on this point. I did not try to change her mind, because her conception of her duty to Einstein was unchangeable, but I made it clear to her that I disliked the use of lawsuits to stop publication of Einstein documents. I had enormous love and respect for Helen as a person, but I never promised that I would support her policy of censorship. I hope and almost believe that, if Helen were now alive and could see with her own eyes that the universal admiration and respect for Einstein have not been diminished by the publication of his intimate letters, she would forgive me.

It is clear to me now that the publication of the intimate letters, even if it is a betrayal of Helen Dukas, is not a betrayal of Einstein. Einstein emerges from this collection of quotes, drawn from many different sources, as a complete and fully rounded human being, a greater and more astonishing figure than the tame philosopher portrayed in Helen's book. Knowledge of the darker side of Einstein's life makes his achievement in science and in public affairs even more miraculous. This book shows him as he was— not a superhuman genius but a human genius, and all the greater for being human.

A few years ago, I had the good luck to be lecturing in Tokyo at the same time as the cosmologist Stephen Hawking. Walking the streets of Tokyo with Hawking in his wheelchair was an amazing experience. I felt as if I were taking a walk through Galilee with Jesus Christ. Everywhere we went, crowds of Japanese silently streamed after us, stretching out their hands to touch Hawking's wheelchair. Hawking enjoyed the spectacle with detached good humor. I was thinking of an account that I had read of Einstein's visit to Japan in 1922. The crowds had streamed after Einstein then as they streamed after Hawking seventy years later. The Japanese people worshiped Einstein as they now worshiped Hawking. They showed exquisite taste in their choice of heroes. Across the barriers of culture and language, they sensed a godlike quality in these two visitors from afar. Somehow they understood that Einstein and Hawking were not just great scientists but great human beings. This book helps to explain why.

Freeman Dyson
Institute for Advanced Study
Princeton, New Jersey, 1996, 2000, 2005, 2010

A (Long) Note about This Final Edition

More than fifteen years have passed since I deliber-
ately began to gather information for the publication of the original edition of *The Quotable Einstein* of
1996. Before that, I had already done so informally
after I began work with the Einstein papers in 1978
at the Institute for Advanced Study in Princeton.
These past years have greatly enriched my life as I
came to know what it's like to be on the other side of
the publishing world—that is, what it's like to be
author rather than editor, to be signing books in
bookstores and for friends, to be reviewed and in-
terviewed. Especially exciting was the year 2005, the
centennial year of Einstein's theory of special relativity and the fiftieth anniversary of his death, when
so many Einstein colleagues and I participated in a
number of international, national, community, and
media events, including the dedication of a long-
awaited statue of Einstein in Princeton.

But now it is time to hang up this particular hat
and wrap things up. This project has been a work in
progress since the beginning, a bit like my ever-
changing but (I hope) improving garden. This fourth
edition is the last one I will compile. I am grateful to
have this last chance to make additions as well as a
number of corrections and clarifications. Perhaps in
a few years an enterprising new editor will have the

energy to continue the project, since there appears to be a bottomless pit of quotable gems to be mined from Einstein's enormous archives.

For this edition, I added three new sections but deleted most of the front- and backmatter found in the earlier editions in order to keep the book a compact size. Readers who are interested in the "extras" of the earlier editions should consult those volumes for the following: a longer chronology of Einstein's life, a family tree, answers to the most common questions about him, excerpts from the FBI's Einstein file, the famous letter to President Roosevelt warning him about the possibility that Germany is building an atom bomb, Johanna Fantova's journal of conversations with Einstein, Helen Dukas's account of Einstein's last days, and a letter to Sigmund Freud from *Why War?* as well as the old introductions, which give some background about my involvement with this project. I hope readers will now enjoy "On and to Children," including Einstein's own two sons; "On Race and Prejudice," which deserves its own section since the publication of Fred Jerome and Rodger Taylor's *Einstein on Race and Racism* has brought new material to light; and just a few verses, limericks, and poems from among the five hundred or so in the archive, all written originally in German. I was able to expand and reorganize the sections covering politics and Jewish themes considerably due to the publication of David Rowe and Robert Schulmann's invaluable and comprehensive book, *Einstein on Politics: His Private and Public Thoughts on Nationalism, Zionism, War, Peace, and the Bomb*, which documents Einstein's political, social, and humanitarian writings. Also of value have been Jürgen Neffe's biography, *Einstein*, meticulously translated into English by Shelley Frisch; Fred Jerome's *Einstein on Israel and Zionism*; and Walter Isaacson's *Einstein*. In addition, the publication of volumes 10 and 12 of the *Collected Papers* has provided yet more primary material. Therefore, I was able to add about four hundred more quotations, bringing the total to around sixteen hundred. The new quotations are prefaced with an asterisk.

While continuing to read about Einstein, I discovered that the versions of his translated writings are continuing to multiply. For example, the reprinted essays in *Ideas and Opinions* and in Nathan and Norden's *Einstein on Peace* are not always faithfully reproduced from the original publications such as *Forum and Century* and the *New York Times Magazine* but were retranslated for these books, for other compilations (such as *Cosmic Religion,* which seems to take snippets from a variety of sources and paraphrase them), and in many biographies thereafter. Because *Ideas and Opinions* is often used as a trusted source, it is not surprising that many of us have been confused about what Einstein *really* said. Perhaps Einstein himself asked that some of his earlier published statements be revised, no longer finding them palatable later in life. It is advisable that scholars, at least, go to the original sources, in the original language, whenever possible, and mention the dates of publication and alternative sources or versions.

Joseph Routh, who became the president of Magdalen College at Oxford over two centuries ago, when asked what precept could serve as a rule of life to an aspiring young man, warned: "You will find it a very good practice always to verify your references, Sir!" In the early editions of this book, I admittedly didn't always heed this sage advice, not being intimately familiar with the vast Einstein literature and believing the sources I had found were trustworthy enough for a general audience. At the same time, I had also warned readers that the original volume was not in fact a scholarly book in the strictest sense. Still, the gist of the quotations themselves has been accurate, with a few exceptions—not too bad when dealing with about sixteen hundred quotations. I've deleted some unverifiable, questionable ones or placed them in the "Attributed to Einstein" section. *This edition therefore supersedes the quotations and sources of the previous editions.*

Furthermore, readers should be aware that published interviews must be taken with a grain of salt since they are filtered by the interviewer and Einstein did not always have a chance to approve them before publication. The same holds true for recollections, conversations, and memoirs, and for anecdotal compilations such as Anita Ehlers's lighthearted *Liebes Hertz!*. I slightly changed a few of the translations that appeared in the earlier editions of this book if I felt the newly found ones were more accurate, and I also added more explanatory material in some notes.

If a quotation can be found in the published volumes of *The Collected Papers of Albert Einstein* (still a work in progress), I included the volume and document number as a source. With this information readers can consult these volumes for further context. Other reliable sources are Rowe and Schulmann's reprinted documents in *Einstein on Politics* and those reprinted in Fred Jerome's books. For items yet unpublished, particularly correspondence, I included an archive number when it was available, as a help to scholars who are able to access the database of the Einstein Archives or the Einstein Papers Project.

Readers of the earlier editions may notice that many of the quotations that were in the original lists in the "Attributed to Einstein" section are no longer in that section and can be found, documented, in the body of the book. The sources of some other popular ones are still undiscovered, and I feel that a good number of them are paraphrases or generalizations of Einstein's thoughts. Many others, though, are completely bogus and continue to be used dishonestly by those who want to use Einstein's name to advance their particular causes.

A word about Einstein's sense of humor is also in order, since humor doesn't always translate well from one language to another. Some, but not all, of his more biting remarks may have been said in jest, tongue-in-cheek, or with a twinkle in his eye. Like most of us, he may have also regretted some of his words later. Once you know Einstein better, you'll also better understand his humor. Furthermore, the reader will note that, through the years, Einstein changed his opinion on a number of topics, as many of us do as we age. So when you read a quotation, be aware that's how he felt at the time he said it, and not necessarily forever after. Such contradictions show that Einstein was not always rigid and narrow-minded but open to new ideas and thoughts as the times demanded, while still trying to remain true to his basic humanitarian values. He was more ironbound in his scientific ideas, though.

Einstein continues and no doubt will forever continue to fascinate both scientists and other admirers around the world. Through his avuncular, genial, and self-effacing image, he manages to exude a charisma that sociologist Max Weber described as "a certain quality of an individual personality, by virtue of which he is set apart from ordinary men and treated as endowed with supernatural, superhuman, or at least specifically exceptional powers or qualities." In this book, however, readers will learn once again that Einstein was all too human and that, for the most part, he is still relevant today.

I thank my editor at Princeton University Press, Ingrid Gnerlich; my production editor, Sara Lerner, for efficiently guiding the manuscript through production; and my eagle-eyed copyeditor, Karen Verde, for their interest and care in helping to prepare this volume for publication. Thanks also go to my former colleagues at Princeton University Press who produced this book and to the many people who, along with their friendly and gracious letters, sent me new quotations and copies of Einstein correspondence I had not seen before. Barbara Wolff of the Einstein Archives in Jerusalem made an exceptional contribution by providing corrections, further sources, and new details that were not familiar to me; she has helped to make this a better book and I'm extremely grateful to her. Osik Moses, an editor at the Einstein Papers Project at Caltech, has been most helpful and was always efficient and quick with her responses, and Diana Buchwald was gracious in giving me access to the archive. As usual, Robert Schulmann provided answers to important questions. Many friends, especially Patrick Lewin, provided encouragement and keen interest in this project. And I again thank Freeman Dyson for his wonderful foreword, which he allowed us to minimally revise in a couple of places. I hope this final edition will serve all readers well.

Claremont, California, January 2010

A Brief Chronology

1879   March 14, Albert Einstein is born in Ulm, Germany.

1880   Family moves to Munich.

1881   November 18, sister Maja is born.

1885   In the fall, enters school and begins violin lessons.

1894   Family moves to Italy, but Albert stays in Munich to finish school. He quits school at the end of the year and joins his family in Italy.

1895   Enters the Aargau Cantonal School in Aarau, Switzerland.

1896   Relinquishes his German citizenship, is graduated from school, and moves to Zurich at the end of October to attend the Swiss Federal Polytechnical Institute (the "Poly"; later the "ETH").

1900   Is graduated from the Polytechnical Institute. Announces he plans to marry fellow student Mileva Marić.

1901   Becomes a Swiss citizen. Seeks employment while tutoring. Begins work on a doctoral dissertation for the University of Zurich.

1902   Probably in January, daughter Lieserl is born out of wedlock to Mileva. June, begins an appointment as Technical Expert at the Patent Office in Bern.

1903   January 6, marries Mileva in Bern, where they take up residence. Lieserl may have been given up for adoption or died, for no mention is made of her after September of this year.

1904   May 14, son Hans Albert is born in Bern.

1905   Einstein's "year of miracles" with respect to his scientific publications.

1906   January 15, receives doctorate from the University of Zurich.

1908   February, becomes a lecturer at the University of Bern.

1909   Is appointed Extraordinary Professor of Physics at the University of Zurich.

1910   July 28, second son, Eduard, is born.

1911   Goes to Prague to teach for a year.

1912   Becomes reacquainted with his divorced cousin Elsa Löwenthal and begins a romantic correspondence with her as his own marriage disintegrates. Accepts appointment as Professor of Theoretical Physics at the Polytechnic Institute (now the ETH) in Zurich.

1913   September, sons Hans Albert and Eduard are baptized as Orthodox Christians near Novi Sad, Hungary (later Yugoslavia, now Serbia), their mother's hometown. Accepts a professorship in Berlin, home of cousin Elsa.

1914   April, arrives in Berlin to assume his new position. Mileva and the children join him but return to Zurich in July because of Einstein's desire to end the marriage.

1916   Publishes "The Origins of the General Theory of Relativity" in *Annalen der Physik*.

1917   October 1, begins directorship of the Kaiser Wilhelm Institute of Physics in Berlin.

1919   February 14, is finally divorced from Mileva. May 29, during a solar eclipse, Sir Arthur Eddington experimentally measures the bending of light and confirms Einstein's predictions; Einstein's fame as a public figure begins. June 2, marries Elsa, who has two daughters at home, Ilse (age 22) and Margot (age 20).

1920   Expressions of anti-Semitism and anti–relativity theory become noticeable among Germans, yet Einstein remains loyal to Germany. Becomes increasingly involved in nonscientific interests, including pacifism and his brand of Zionism.

1921   April and May, makes first trip to the United States. Accompanies Chaim Weizmann on U.S. fund-raising tour on behalf of Hebrew University of Jerusalem. Delivers four lectures on relativity theory at Princeton University.

1922   October through December, takes trip to the Far East. November, while in Shanghai, learns that he has won the 1921 Nobel Prize in physics.

1923   Visits Palestine and Spain.

1925   Travels to South America. In solidarity with Gandhi, signs a manifesto against compulsory military service. Becomes an ardent pacifist.

1928   April, Helen Dukas is hired as his secretary and remains with him as secretary and housekeeper for the rest of his life.

1930   December, visits New York and Cuba, then stays (until March 1931) at the California Institute of Technology (Caltech), in Pasadena.

1931   Visits Oxford in May to deliver the Rhodes Lectures. December, en route to Pasadena again.

1932   January–March, at Caltech. Returns to Berlin. December, takes another trip to the United States.

1933   January, Nazis come to power in Germany. Gives up German citizenship (remains a Swiss citizen) and does not return to Germany. Instead, from the United States, goes to Belgium with Elsa and sets up temporary residence at Coq sur Mer.

Takes trips to Oxford, where he delivers the Herbert Spencer Lecture in June, and Switzerland, where he makes his final visit to son Eduard in a psychiatric hospital. Early October, leaves Europe for Princeton, New Jersey, to begin professorship at the Institute for Advanced Study.

1936   December 20, Elsa dies after a long battle with heart and kidney disease.

1939   August 2, signs famous letter to President Roosevelt on the military implications of atomic energy, which leads to the Manhattan Project.

1940   Becomes U.S. citizen.

1945   Retires officially from the faculty of the Institute for Advanced Study.

1948   August 4, Mileva dies in Zurich.

1950   March 18, signs his last will. His literary estate (the archive) is to be transferred to the Hebrew University of Jerusalem at a time determined by his trustees.

1952   Is offered the presidency of Israel, which he declines.

1955   April 11, writes last signed letter, to Bertrand Russell, agreeing to sign a joint manifesto urging all nations to renounce nuclear weapons. April 13, aneurysm ruptures. April 15, enters Princeton Hospital. April 18, Albert Einstein dies at 1:15 A.M. of a ruptured arteriosclerotic aneurysm of the abdominal aorta.

The Ultimate *Quotable*
EINSTEIN

On Einstein Himself

A happy man is too satisfied with the present to think too much about the future.

Written at age seventeen (September 18, 1896) for a school essay in French entitled "My Future Plans." CPAE, Vol. 1, Doc. 22

Strenuous intellectual work and the study of God's Nature are the angels that will lead me through all the troubles of this life with consolation, strength, and uncompromising rigor.

To Pauline Winteler, mother of Einstein's girlfriend Marie, May (?) 1897. CPAE, Vol. 1, Doc. 34

*In many a lucid moment I appear to myself as an ostrich who buries his head in the desert sand so as not to perceive a danger. One creates a small world for oneself and . . . one feels miraculously great and important, just like a mole in its self-dug hole.

Ibid.

*I know this sort of animal personally, from my own experience, as I am one of them myself. Not too much should be expected of them. . . . Today we are sullen, tomorrow high-spirited, after tomorrow cold, then again irritated and half-sick of life—not to mention unfaithfulness, ingratitude, and selfishness.

To friend Julia Niggli, ca. August 6, 1899, after she asked him his opinion about her relationship with an older man. CPAE, Vol. 1, Doc. 51

I decided the following about our future: I will look for a position immediately, no matter how modest it is. My scientific goals and my personal vanity will not prevent me from accepting even the most subordinate position.

To future wife Mileva Marić, ca. July 7, 1901, while having difficulty finding his first job. *CPAE*, Vol. 1, Doc. 114

In living through this "great epoch," it is difficult to reconcile oneself to the fact that one belongs to that mad, degenerate species that boasts of its free will. How I wish that somewhere there existed an island for those who are wise and of good will! In such a place even I should be an ardent patriot!

To Paul Ehrenfest, early December 1914. *CPAE*, Vol. 8, Doc. 39

Do not feel sorry for me. Despite terrible appearances, my life goes on in full harmony; I am entirely devoted to reflection. I resemble a farsighted man who is charmed by the vast horizon and who is disturbed by the foreground only when an opaque object obstructs his view.

To Helene Savić, September 8, 1916, after separation from his family. In Popović, ed., *In Albert's Shadow*, 110. *CPAE*, Vol. 8, Doc. 258

I very rarely think in words at all. A thought comes, and I may try to express it in words afterwards.

From a conversation with psychologist Max Wertheimer in 1916. In Wertheimer, *Productive Thinking* (New York: Harper, 1945), footnote on p. 184

I have come to know the mutability of all human relationships and have learned to insulate myself against both heat and cold so that a temperature balance is fairly well assured.

To Heinrich Zangger, March 10, 1917. *CPAE*, Vol. 8, Doc. 309

I am by heritage a Jew, by citizenship a Swiss, and by disposition a human being, and *only* a human being, without any special attachment to any state or national entity whatsoever.

To Adolf Kneser, June 7, 1918. *CPAE*, Vol. 8, Doc. 560

I was originally supposed to become an engineer, but the thought of having to expend my creative energy on things that make practical everyday life even more refined, with a loathsome capital gain as the goal, was unbearable to me.

To Heinrich Zangger, ca. August 1918. *CPAE*, Vol. 8, Doc. 597

I lack any sentiment of the sort; all I have is a sense of duty toward all people and an attachment to those with whom I have become intimate.

To Heinrich Zangger, June 1, 1919, regarding his lack of attachment to any particular place, as, for example, physicist Max Planck had to Germany. *CPAE*, Vol. 9, Doc. 52

I also had little inclination for history [in school].
But I think it had more to do with the method of instruction than with the subject itself.

> To sons Hans Albert and Eduard, June 13, 1919. *CPAE*,
> Vol. 9, Doc. 60

I have not yet eaten enough of the fruit of the Tree of
Knowledge, though in my profession I am obliged
to feed on it regularly.

> To Max Born, November 9, 1919. In Born, *Born-Einstein Let-
> ters*, 16; *CPAE*, Vol. 9, Doc. 162

By an application of the theory of relativity to the
taste of readers, to-day in Germany I am called a
German man of science, and in England I am represented as a Swiss Jew. If I come to be represented as
a *bête noire*, the descriptions will be reversed, and I
shall become a Swiss Jew for the Germans and a
German man of science for the English!

> To *The Times* (London), November 28, 1919, 13–14, written
> at the request of the newspaper. Also referred to in a letter
> to Paul Ehrenfest, December 4, 1919. See also the quotation
> of April 6, 1922, below. *CPAE*, Vol. 7, Doc. 26

Another funny thing is that I myself count everywhere as a Bolshevist, God knows why; perhaps because I do not take all that slop in the *Berliner Tage-
blatt* as milk and honey.

> To Heinrich Zangger, December 15 or 22, 1919. *CPAE*, Vol. 9,
> Doc. 217

With fame I become more and more stupid, which of course is a very common phenomenon.

To Heinrich Zangger, December 24, 1919. *CPAE*, Vol. 9, Doc. 233

Since the light deflection result became public, such a cult has been made out of me that I feel like a pagan idol. But this, too, God willing, will pass.

To Heinrich Zangger, January 3, 1920. *CPAE*, Vol. 9, Doc. 242. Einstein had even been asked to give a three-week "performance" at the London Palladium to explain relativity.

I do know that kind fate allowed me to find a couple of nice ideas after many years of feverish labor.

To Dutch physicist H. A. Lorentz, January 19, 1920. *CPAE*, Vol. 9, Doc. 265

An awareness of my limitations pervades me all the more keenly in recent times because my faculties have been quite overrated since a few consequences of general relativity theory have stood the test.

Ibid.

I am being so terribly deluged with inquiries, invitations, and requests that at night I dream I am burning in hell and the mailman is the devil and is continually yelling at me, hurling a fresh bundle of letters at my head because I still haven't answered the old ones.

To Ludwig Hopf, February 2, 1920. CPAE, Vol. 9, Doc. 295

My father's ashes lie in Milan. I buried my mother here [Berlin] only a few days ago. I myself have journeyed to and fro continuously—a stranger everywhere. My children are in Switzerland. . . . A person like me has as his ideal to be at home anywhere with his near and dear ones.

To Max Born, March 3, 1920. In Born, *Born-Einstein Letters*, 25. *CPAE*, Vol. 9, Doc. 337

The teaching faculty in elementary school was liberal and did not make any denominational distinctions. Among the *Gymnasium* teachers there were a few anti-Semites. Among the children, anti-Semitism was alive especially in elementary school. It was based on conspicuous racial characteristics and on impressions left from the lessons on religion. Active attacks and verbal abuse on the way to and from school were frequent but usually not all that serious. They sufficed, however, to establish an acute feeling of alienation already in childhood.

To Paul Nathan, political editor of the *Berliner Tageblatt*, for an article on anti-Semitism, April 3, 1920. *CPAE*, Vol. 9, Doc. 366

I will always fondly recall the hours spent in your home, including the pearls of Persian wisdom with which I became acquainted through your hospitality and your work. As an Oriental by blood, I feel they are especially meaningful to me.

> To Friedrich Rosen, German envoy in The Hague, May 1920. Rosen had apparently been posted in Persia at one time and edited a collection of Persian stories. Einstein Archives 9-492

It also pleases me that it is still possible, even today, to be treated as an internationally minded person without being compartmentalized into one of the two big drawers.

> To H. A. Lorentz, June 15, 1920. The "two big drawers" at the time were the pro–Central Powers and the pro-Allies. *CPAE*, Vol. 10, Doc. 56

*Don't be too hard on me. Everyone has to sacrifice at the altar of stupidity from time to time, to please the Deity and the human race. And this I have done thoroughly with my article.

> To Max and Hedi Born, September 9, 1920, downplaying criticism for an article he wrote. In Born, *Born-Einstein Letters*, 34. *CPAE*, Vol. 7, Doc. 45

Like the man in the fairytale who turned everything he touched into gold, so with me everything is turned into newspaper clamor.

> Ibid. To his friend Paul Ehrenfest he wrote ten years later, on March 21, 1930, "With me, every peep becomes a trumpet solo" (Einstein Archives 10-212).

Personally, I experience the greatest degree of pleasure in having contact with works of art. They furnish me with happy feelings of an intensity that I cannot derive from other sources.

> 1920. Quoted by Moszkowski, *Conversations with Einstein*, 184. Here, according to the context, Einstein refers only to literature.

*I do not care to speak about my work. The sculptor, the artists, the musician, the scientist work because they love their work. Fame and honor are secondary. My work is my life, and when I find the truth I proclaim it. . . . Opposition does not affect my work.

> Quoted in *New York Call*, May 31, 1921, 2. See also Illy, *Albert Meets America*, 312

To be called to account publicly for what others have said in your name, when you cannot defend yourself, is a sad situation indeed.

> From "Einstein and the Interviewers," August 1921. Einstein Archives 21-047

If my theory of relativity is proven successful, Germany will claim me as a German and France will declare that I am a citizen of the world. Should my theory prove untrue, France will say that I am a German and Germany will declare that I am a Jew.

> From an address to the French Philosophical Society at the Sorbonne, April 6, 1922. See also French press clipping,

April 7, 1922, Einstein Archives 36-378; and *Berliner Tage-blatt*, April 8, 1922. Einstein Archives 79-535

When a blind beetle crawls over the surface of a curved branch, it doesn't notice that the track it has covered is indeed curved. I was lucky enough to notice what the beetle didn't notice.

In answer to his son Eduard's question about why he is so famous, 1922. Quoted in Flückiger, *Albert Einstein in Bern*, and Grüning, *Ein Haus für Albert Einstein*, 498

Now I am sitting peacefully in Holland after being told that certain people in Germany have it in for me as a "Jewish saint." In Stuttgart there was even a poster in which I appeared in first place among the richest Jews.

To sons Hans Albert and Eduard, November 24, 1923. Einstein Archives 75-627

Of all the communities available to us, there is not one I would want to devote myself to except for the society of the true searchers, which has very few living members at any one time.

To Max and Hedwig Born, April 29, 1924. In Born, *Born-Einstein Letters*, 79. Einstein Archives 8-176

[I] must seek in the stars that which was denied [to me] on Earth.

To his secretary Betty Neumann, 1924, with whom he had fallen in love while married to Elsa, upon ending

his relationship with her. She was the niece of his friend
Hans Muehsam. See Pais, *Subtle Is the Lord*, 320; and Fölsing, *Albert Einstein*, 548

Imagination is more important than knowledge.
Knowledge is limited. Imagination encircles the
world.

In answer to the question, "Do you trust more to your
imagination than to your knowledge?" From interview
with G. S. Viereck, "What Life Means to Einstein," *Saturday
Evening Post*, October 26, 1929; reprinted in Viereck,
Glimpses of the Great, 447

My own career was undoubtedly determined not by
my own will, but by various factors over which I
have no control, primarily those mysterious glands
in which nature prepares the very essence of life.

In a discussion on free will and determinism. Ibid. Reprinted in Viereck, *Glimpses of the Great*, 442

To punish me for my contempt of authority, Fate has
made me an authority myself.

Aphorism for a friend, September 18, 1930. Quoted in Hoffmann, *Albert Einstein: Creator and Rebel*, 24. Einstein Archives 36-598

I am an artist's model.

As recalled and noted by Herbert Samuel, who asked him
his occupation, reflecting Einstein's feeling that he was
constantly posing for sculptures and paintings, October 31,
1930. Einstein Archives 21-006. The photographer Philippe

Halsmann's version is a bit different: An elderly woman on a bus told Einstein she must have seen his picture somewhere because he looked familiar, and Einstein responded, "I am a photographer's model." See Halsmann, letter to editors, *New York Review of Books*, May 26, 1966

I have never looked upon ease and happiness as ends in themselves—such an ethical basis I call the ideal of a pigsty. . . . The ideals which have always shone before me and filled me with the joy of living are goodness, beauty, and truth. To make a goal of comfort or happiness has never appealed to me.

From "What I Believe," *Forum and Century* 84 (1930), 193–194. See also Rowe and Schulmann, *Einstein on Politics*, 226, for background information and the whole essay. This and other passages from the essay have been variously translated elsewhere. For this edition of the book, I am consistently using the versions in *Forum and Century*.

*Possessions, outward success, publicity, luxury—to me these have always been contemptible. I believe that a simple and unassuming life is best for . . . the body and mind.

Ibid.

*My passionate interest in social justice and social responsibility has always stood in curious contrast to a marked lack of desire for direct association with men and women. I am a horse for single harness, not cut out for tandem or team work. I have never belonged wholeheartedly to country or state, to my circle of friends, or even to my own family. These ties have always been accompanied by a vague aloofness, and the wish to withdraw into myself increases with the years.

Ibid.

Many times a day I realize how much my outer and inner life is based upon the labors of my fellow men, both living and dead, and how much I must exert myself in order to give in return as much as I have received.

Ibid.

It is an irony of fate that I myself have been the recipient of excessive admiration and reverence from my fellow-beings, through no fault or merit of my own.

Ibid.

Professor Einstein begs you to treat your publications for the time being as if he were already dead.

Written on Einstein's behalf by his secretary, Helen Dukas, March 1931, after he was besieged by one too many manuscripts. Einstein Archives 46-487

It strikes me as unfair, and even in bad taste, to select a few individuals for boundless admiration, attributing superhuman powers of mind and character to them. This has been my fate, and the contrast between the popular assessment of my powers and achievements and the reality is simply grotesque.

From "Impressions of the U.S.A.," ca. 1931, reprinted in Rowe and Schulmann, *Einstein on Politics*, 242–246. Einstein Archives 28-168

Although I try to be universal in thought, I am European by instinct and inclination.

Quoted in *Daily Express* (London), September 11, 1933. Also in Holton, *Advancement of Science*, 126

People flatter me as long as I'm of use to them. But when I try to serve goals with which they are in disagreement, they immediately turn to abuse and calumny in defense of their interests.

To an unidentified pacifist, 1932. Einstein Archives 28-191

I suffered at the hands of my teachers a similar treatment; they disliked me for my independence and passed me over when they wanted assistants. (I must admit, though, that I was somewhat less of a model student than you.)

To a young girl, Irene Freuder, November 20, 1932. Reprinted as "Education and Educators," in *Ideas and Opinions*, 56. Einstein Archives 28-221

My life is a simple thing that would interest no one. It is a known fact that I was born, and that is all that is necessary.

To Princeton High School reporter Henry Russo, quoted in *The Tower*, April 13, 1935

As a boy of twelve years making my acquaintance with elementary mathematics, I was thrilled in seeing that it was possible to find out truth by reasoning alone, without the help of any outside experience. . . . I became more and more convinced that even nature could be understood as a relatively simple mathematical structure.

Ibid.

Arrows of hate have been aimed at me too, but they have never hit me, because somehow they belonged to another world with which I have no connection whatsoever.

From a statement written for Georges Schreiber's *Portraits and Self-Portraits* (Boston: Houghton Mifflin, 1936). Reprinted in *Out of My Later Years*, 13. Einstein Archives 28-332

I have settled down splendidly here: I hibernate like a bear in its cave, and really feel more at home than ever before in all my varied existence. This bearishness has been accentuated still further because of the death of my mate, who was more attached to human beings than I.

To Max Born, early 1937, after the death of Einstein's wife, Elsa. In Born, *Born-Einstein Letters*, 125. Einstein Archives 8-199

I wouldn't want to live if I did not have my work. . . . In any case, it's good that I'm already old and personally don't have to count on a prolonged future.

> To close friend Michele Besso, October 10, 1938, reflecting on Hitler's rise to power. Einstein Archives 7-376

I firmly believe that love [of a subject or hobby] is a better teacher than a sense of duty—at least for me.

> In draft of a letter to Philipp Frank, 1940. Einstein Archives 71-191

*I have never given my name for commercial use even in cases where no misleading of the public was involved as it would be in your case. I, therefore, forbid you to use my name in any way.

> To Marvin Ruebush, who had asked Einstein for permission to use his name in promoting a cure for stomach aches, May 22, 1942. Einstein Archives 56-066

Why is it that nobody understands me, yet everybody likes me?

> From an interview, *New York Times*, March 12, 1944

I do not like to state an opinion on a matter unless I know the precise facts.

> From an interview with Richard J. Lewis, *New York Times*, August 12, 1945, 29:3, on declining to comment on Germany's progress on the atom bomb

I never think of the future. It comes soon enough.

> Aphorism, 1945–46. According to the Oxford Dictionary of
> Humorous Quotations (2d ed., 2001), this quotation came
> from an interview on the ship *Belgenland* in December 1930;
> perhaps it was recalled later and inserted into the
> archives under the later date. Einstein Archives 36-570

The development of this thought world (*Gedanken-welt*) is in a certain sense a continuous flight from "wonder." A wonder of such nature I experienced as a child of four or five years, when my father showed me a compass.

> Written in 1946 for "Autobiographical Notes," 9

My intuition was not strong enough in the field of mathematics in order to differentiate clearly the fundamentally important . . . from the rest of the more or less dispensable erudition. Beyond this, however, my interest in the knowledge of nature was also unqualifiedly stronger. . . . In this field I soon learned to scent out that which was able to lead to fundamentals and to turn aside . . . from the multitude of things which clutter up the mind and divert it from the essential.

> Ibid., 15–17

The essential in the being of a man of my type lies precisely in *what* he thinks and *how* he thinks, not in what he does or suffers.

> Ibid., 33

There have already been published by the bucketsful such brazen lies and utter fictions about me that I would long since have gone to my grave if I had allowed myself to pay attention to them.

To the writer Max Brod, February 22, 1949. Einstein Archives 34-066.1

*I lack influence [at the Institute for Advanced Study], as I am generally regarded as a sort of petrified object, rendered blind and deaf by the years. I find this role not too distasteful, as it corresponds fairly well with my temperament.

To Max and Hedi Born, April 12, 1949. In Born, *Born-Einstein Letters*, 178–179. (Similar to "My fame begins outside of Princeton. My word counts for little in Fine Hall," as quoted by Infeld in *Quest*, 302.) Einstein Archives 8-223

*I simply enjoy giving more than receiving in every respect, to not take myself nor the doings of the masses seriously, am not ashamed of my weaknesses and vices, and naturally take things as they come with equanimity and humor. Many people are like this, and I really cannot understand why I have been made into a kind of idol.

Ibid., in reply to Max Born's question on Einstein's attitude toward a simple life

My scientific work is motivated by an irresistible longing to understand the secrets of nature and by no other feelings. My love for justice and the striving to contribute toward the improvement of human conditions are quite independent from my scientific interests.

> To F. Lentz, August 20, 1949, in answer to a letter asking Einstein about his scientific motivation. Einstein Archives 58-418

I have no special talents. I am only passionately curious.

> To Carl Seelig, March 11, 1952. Einstein Archives 39-013

I'm doing just fine, considering that I have triumphantly survived Nazism and two wives.

> To Jakob Ehrat, May 12, 1952. Einstein Archives 59-554

It is a strange thing to be so widely known, and yet to be so lonely. But it is a fact that this kind of popularity . . . is forcing its victim into a defensive position which leads to isolation.

> To E. Marangoni, October 1, 1952. Einstein Archives 60-406

All my life I have dealt with objective matters; hence I lack both the natural aptitude and the experience to deal properly with people and to carry out official functions.

> Statement to Abba Eban, Israeli ambassador to the United States, November 18, 1952, in turning down the presidency of Israel after Chaim Weizmann's death. Einstein Archives 28-943

*I myself have certainly found satisfaction in my efforts, but I would not consider it sensible to defend the results of my work as being my own "property," like some old miser might defend the few pennies he had laboriously scraped together.

To Max Born, October 12, 1953. In Born, *Einstein-Born Letters*, 195. Einstein Archives 8-231

I'm a magnet for all the crackpots in the world, but they are of interest to me, too. A favorite pastime of mine is to reconstruct their thinking processes. I feel genuinely sorry for them, that's why I try to help them.

Quoted by Fantova, "Conversations with Einstein," October 15, 1953

In the past it never occurred to me that every casual remark of mine would be snatched up and recorded. Otherwise I would have crept further into my shell.

To Carl Seelig, October 25, 1953. Einstein Archives 39-053

During the First World War, when I was thirty-five years old and traveled from Germany to Switzerland, I was stopped at the border and asked for my name. I had to hesitate before I remembered it. I have always had a bad memory.

Quoted by Fantova, "Conversations with Einstein," November 7, 1953

I was supposed to be named Abraham after my grandfather. But that was too Jewish for my parents, so they made use of the "A" and named me Albert.

Ibid., December 5, 1953

All manner of fable is being attached to my personality, and there is no end to the number of ingeniously devised tales. All the more do I appreciate and respect what is truly sincere.

To Queen Elisabeth of Belgium, March 28, 1954. Einstein Archives 32-410

Today Mr. Berks has shown me the bust he made of me. I admire the bust highly as a portrait and not less as a work of art and as a characterization of mental personality.

From a signed statement written in English, April 15, 1954. Robert Berks is the sculptor who created the statue of Einstein in front of the National Academy of Science in Washington, D.C. The bust was used as a model for the statue. The bust itself, donated by the sculptor, was placed in front of Borough Hall, Princeton, New Jersey, in April 2005. (Statement, of which he gave me a copy, is in possession of Mr. Berks.)

It is true that my parents were worried because I began to speak fairly late, so that they even consulted a doctor. I can't say how old I was—but surely not less than three.

To Sybille Blinoff, May 21, 1954. Einstein Archives 59-261. In her biography of Einstein, Einstein's sister, Maja, put his age at two and a half; see *CPAE*, Vol. 1, lvii

I'm not the kind of snob or exhibitionist that you take me to be and furthermore have nothing of value to say of immediate concern, as you seem to assume.

In reply to a letter, May 27, 1954, asking Einstein to send a message to a new museum in Chile, to be put on display for others to admire. Einstein Archives 60-624

It is quite curious, even abnormal, that, with your superficial knowledge about the subject, you are so confident in your judgment. I regret that I cannot spare the time to occupy myself with dilettantes.

To dentist G. Lebau, who claimed he had a better theory of relativity, July 10, 1954. The dentist returned Einstein's letter with a note written at the bottom: "I am thirty years old; it takes time to learn humility." Einstein Archives 60-226

I never read what anyone writes about me—they are mostly lies from the newspapers that are always repeated. . . . The only exception has been the Swiss man, [Carl] Seelig; he is very nice and did a good job. I didn't read his book, either, but Dukas read some parts of it to me.

Quoted by Fantova, "Conversations with Einstein," September 13, 1954

If I would be a young man again and had to decide how to make my living, I would not try to become a scientist or scholar or teacher. I would rather choose to be a plumber or a peddler, in the hope of finding that modest degree of independence still available under present circumstances.

> To the editor, *The Reporter* 11, no. 9 (November 18, 1954). See Rowe and Schulmann, *Einstein* on Politics, 485–486. Said in response to the McCarthy-era witch hunt of intellectuals. He felt that science at its best should be a hobby and that one should make a living at something else (see Straus, "Reminiscences," in Holton and Elkana, *Albert Einstein: Historical and Cultural Perspectives*, 421). A plumber, Stanley Murray, replied to Einstein on November 11: "Since my ambition has always been to be a scholar and yours seems to be a plumber, I suggest that as a team we would be tremendously successful. We can then be possessed of both knowledge and independence" (Rosenkranz, *Einstein Scrapbook*, 82–83). At other times, Einstein allegedly also claimed that he would choose to be a musician, and suggested the job of lighthouse keeper to young scientists in a speech in the Royal Albert Hall in London in 1933 (Nathan and Norden, *Einstein on Peace*, 238).

*In the present circumstances, the only profession I would choose would be one where earning a living had nothing to do with the search for knowledge.

> To Max Born, January 17, 1955. See Born, *Born-Einstein Letters*, 227. Einstein Archives 8-246

Only in mathematics and physics was I, through self-study, far beyond the school curriculum, and also with regard to philosophy as it was taught in the school curriculum.

To Henry Kollin, February 1955. Quoted in Hoffmann, *Albert Einstein: Creator and Rebel*, 20. Einstein Archives 60-046

The only way to escape the corruptible effect of praise is to go on working.

Quoted by Lincoln Barnett, "On His Centennial, the Spirit of Einstein Abides in Princeton," Smithsonian, February 1979, 74

God gave me the stubbornness of a mule and a fairly keen scent.

As recalled by Ernst Straus. Quoted in Seelig, *Helle Zeit, dunkle Zeit*, 72

The ordinary adult never gives a thought to space-time problems. . . . I, on the contrary, developed so slowly that I did not begin to wonder about space and time until I was an adult. I then delved more deeply into the problem than any other adult or child would have done.

As recalled by Nobel laureate James Franck, on Einstein's belief that it is usually children, not adults, who reflect on space-time problems. Quoted in Seelig, *Albert Einstein und die Schweiz*, 73

When I was young, all I wanted and expected from life was to sit quietly in some corner doing my work

without the public paying attention to me. And now
see what has become of me.

Quoted in Hoffmann, *Albert Einstein: Creator and Rebel*, 4

When I examine myself and my methods of thought,
I come close to the conclusion that the gift of imagination has meant more to me than my talent for ab-
sorbing absolute knowledge.

Similar to "Imagination is more important than knowledge" (1929), quoted above. Recalled by a friend on the one
hundredth anniversary of Einstein's birth, celebrated February 18, 1979. Quoted in Ryan, *Einstein and the Humanities*, 125

I have never obtained any ethical values from my
scientific work.

As recalled by Manfred Clynes. Quoted in Michelmore,
Einstein: Profile of the Man, 251

Many things which go under my name are badly
translated from the German or are invented by other
people.

To George Seldes, compiler of *The Great Quotations* (1960),
cited in Kantha, *An Einstein Dictionary*, 175

I hate my pictures. Look at my face. If it weren't for
this [his mustache], I'd look like a woman!

Said to photographer Alan Richards sometime during the
last ten years of his life. Quoted by Richards, "Reminiscences," in *Einstein as I Knew Him* (unnumbered pages)

You're the first person in years who has told me what you really think of me.

> To an eighteen-month-old baby boy who screamed upon being introduced to Einstein. Quoted in ibid.

I have finished my task here.

> Said as he was dying. Einstein Archives 39-095. Taken from biographer Carl Seelig's account; he may have heard it from Einstein's secretary Helen Dukas or stepdaughter Margot Einstein.

On and to His Family

According to Einstein, though his marriage to Mileva, a Serbian woman, lasted for seventeen years, he never really knew her. He recalled that he had married her primarily "from a sense of duty," possibly because she had given birth to their illegitimate child. "I had, with an inner resistance, embarked on something that simply exceeded my strength." They had met at the Swiss Federal Polytechnical Institute, where both were physics students; he was eighteen and she was twenty-two. At the time of their marriage about five years later, he was not aware that mental illness was hereditary on Mileva's mother's side of the family. Mileva herself was often depressed and her sister, Zorka, was schizophrenic. Still, she was a warm and caring woman and highly intelligent, and had much to cope with throughout life. Because of her inability to accept her pending divorce, Einstein's often insensitive treatment of her, and the decision not to bring their illegitimate daughter, Lieserl, to live with them, Mileva became bitter, sometimes causing difficulties in Einstein's relationship with his two sons during their separation. The many letters he wrote to them, especially to Hans Albert, show that he tried to remain close to them during their childhood and that he regarded them warmly and with care and concern. He also eventually conceded that Mileva was a good mother. (See *CPAE*, Vol. 8, for these letters as well as letters to Mileva in which the couple tries to deal with its financial and parenting difficulties after the separation. See also Popović, ed., *In Albert's Shadow*.)

Still, these tragic circumstances of their separation, according to Einstein, left their mark on him into his old age and may have amplified his deep involvement in activities of an impersonal nature. See letters to his biographer Carl Seelig, March 26 and May 5, 1952; Einstein Archives 39-016 and 39-020

Mama threw herself on the bed, buried her head in the pillow, and wept like a child. After regaining her composure, she immediately shifted to a desperate attack: "You're ruining your future and destroying your opportunities." "No decent family would want her." "If she becomes pregnant, you'll be in a real mess." With this outburst, which was preceded by many others, I finally lost my patience.

To Mileva, July 29, 1900, after telling his mother that he and Mileva planned to marry; they did not marry until January 6, 1903. *The Love Letters*, 19; *CPAE*, Vol. 8, Doc. 68

I long terribly for a letter from my beloved witch. I find it hard to believe that we will be separated for so much longer—only now do I see how much in love with you I am! Pamper yourself, so you will become a radiant little sweetheart and as wild as a street urchin!

To Mileva, August 1, 1900. *The Love Letters*, 21; *CPAE*, Vol. 1, Doc. 69

When you're not with me, I feel as though I'm not complete. When I'm sitting, I want to go away; when

I go away, I'd rather be home; when I'm talking with people, I'd rather be studying; when I study, I can't sit still and concentrate; and when I go to sleep, I'm not satisfied with the way the day has passed.

To Mileva, August 6, 1900. *The Love Letters*, 23–24; *CPAE*, Vol. 1, Doc. 70

How was I ever able to live alone, my little everything? Without you I have no self-confidence, no passion for work, and no enjoyment of life—in short, without you, my life is a void.

To Mileva, ca. August 14, 1900. *The Love Letters*, 26; *CPAE*, Vol. 1, Doc. 72

My parents are very concerned about my love for you. . . . They cry for me almost as if I had already died. Again and again they complain that I brought misfortune on myself by my devotion to you.

To Mileva, August–September 1900. *The Love Letters*, 29; *CPAE*, Vol. 1, Doc. 74

Without the thought of you, I would no longer want to live among this sorry herd of humans. But having you makes me proud, and the thought of you makes me happy. I will be doubly happy when I can press you to my heart once again and see those loving eyes shine for me alone, and when I can kiss that sweet mouth that trembles for me only.

Ibid.

I am also looking forward to working on our new studies. You must continue with your research—how proud I will be to have a little Ph.D. for a sweetheart while I remain a totally ordinary person!

> To Mileva, September 13, 1900. *The Love Letters*, 32; *CPAE*, Vol. 1, Doc. 5

Shall I look around for possible jobs for you [in Zurich]? I think I'll try to find some tutoring positions that I can later turn over to you. Or do you have something else in mind? . . . No matter what happens, we'll have the most wonderful life imaginable.

> To Mileva, September 19, 1900. *The Love Letters*, 33; *CPAE*, Vol. 1, Doc. 76

I am so lucky to have found you—a creature who is my equal, and who is as strong and independent as I am.

> To Mileva, October 3, 1900. *The Love Letters*, 36; *CPAE*, Vol. 1, Doc. 79

How happy and proud I will be when the two of us together have brought our work on relative motion to a triumphant end!

> To Mileva, March 27, 1901. *The Love Letters*, 29; *CPAE*, Vol. 1, Doc. 94. This sentence has led some to believe that Mileva was equally responsible for the theory of relativity.

You'll see for yourself how pleasant and cheerful I've become and how all of my scowling is a thing of the past. And I love you so much again! It was only because of nervousness that I was so mean to you . . . and I'm longing so much to see you again.

To Mileva, April 30, 1901. *The Love Letters*, 46; *CPAE*, Vol. 1, Doc. 102

If only I could give you some of my happiness so you would never be sad and depressed again.

To Mileva, May 9, 1901. *The Love Letters*, 51; *CPAE*, Vol. 1, Doc. 106

My wife is coming to Berlin with very mixed feelings because she is afraid of the relatives, probably mostly of you. . . . But you and I can be very happy with each other without her having to be hurt. You can't take away from her something she doesn't have [i.e., his love].

To newfound love, cousin Elsa Löwenthal, August 1913. *CPAE*, Vol. 5, Doc. 465

The situation in my house is ghostlier than ever: icy silence.

To Elsa, October 16, 1913. *CPAE*, Vol. 5, Doc. 478

Do you think it's so easy to get a divorce when one has no proof of the other party's guilt? . . . I am treating my wife like an employee whom I can't fire. I have my own bedroom and avoid being with her. . . . I don't know why you're so terribly upset by all of this. I'm absolutely my own master . . . as well as my own wife.

> To Elsa, before December 2, 1913. *CPAE*, Vol. 5, Doc. 488

[My wife, Mileva] is an unfriendly, humorless creature who gets nothing out of life and who, by her mere presence, extinguishes other people's joy of living.

> To Elsa, after December 2, 1913. *CPAE*, Vol. 5, Doc. 489

My wife whines incessantly to me about Berlin and her fear of the relatives. . . . My mother is good-natured, but she is a really fiendish mother-in-law. When she stays with us, the air is full of dynamite. . . . But both are to be blamed for their miserable relationship. . . . No wonder that my scientific life thrives under these circumstances: it lifts me impersonally from the vale of tears into a more peaceful atmosphere.

> To Elsa, after December 21, 1913. *CPAE*, Vol. 5, Doc. 497

*He had kind of a relationship with my wife, which no one can hold against them.

> To Heinrich Zangger, June 27, 1914. Einstein presumed Mileva had an affair with Vladimir Varićak, a professor of mathematics at the University of Zagreb, who made two important discoveries in the theory of relativity, which were cited by Wolfgang Pauli in his review paper on relativity. *CPAE*, Vol. 8, Doc. 34a, embedded in Vol. 10

(A) You will see to it that (1) my clothes and laundry are kept in good order; (2) I will be served three meals regularly in my room; (3) my bedroom and study are kept tidy, and especially that my desk is left for my use only. (B) You will relinquish all personal relations with me insofar as they are not completely necessary for social reasons. Particularly, you will forgo my (1) staying at home with you; (2) going out or traveling with you. (C) You will obey the following points in your relations with me: (1) you will not expect any tenderness from me, nor will you offer any suggestions to me; (2) you will stop talking to me about something if I request it; (3) you will leave my bedroom or study without any backtalk if I request it. (D) You will undertake not to belittle me in front of our children, either through words or behavior.

> Memorandum to Mileva, ca. July 18, 1914, listing the conditions under which he would agree to continue to live with her in Berlin. At first she accepted the conditions, but then left Berlin with the children at the end of July. *CPAE*, Vol. 8, Doc. 22

I don't want to lose the children, and I don't want them to lose me. . . . After everything that has happened, a friendly relationship with you is out of the question. We shall have a considerate and business-like relationship. All personal things must be kept to a minimum. . . . I don't expect I'll ask you for a divorce but only want you to stay in Switzerland with the children . . . and send me news of my precious boys every two weeks. . . . In return, I assure you of proper comportment on my part, such as I would exercise toward any unrelated woman.

> To Mileva, ca. July 18, 1914, on his offer to continue their marriage after his move to Berlin, to which in the end she did not agree. *CPAE*, Vol. 8, Doc. 23

I came to realize that living with the children is no blessing if the wife stands in the way.

> To Elsa, July 26, 1914. *CPAE*, Vol. 8, Doc. 26

I may see my children only on neutral ground, not in our [future] home. This is justified because it is not right to have the children see their father with a woman other than their own mother.

> To Elsa, after July 26, 1914. *CPAE*, Vol. 8, Doc. 27

How much I look forward to the quiet evenings we'll be able to spend chatting alone, and to all the peaceful shared experiences still ahead of us! Now, after all my deliberations and work I'll find a precious little wife at home who receives me with cheer and contentment. . . . It wasn't her [Mileva's] ugliness, but her obstinacy, inflexibility, stubbornness, and insensitivity that prevented harmony between us.

> To Elsa, July 30, 1914. *CPAE*, Vol. 8, Doc. 30

There are reasons why I could not endure being with this woman any longer, despite the tender love that ties me to the children.

To Heinrich Zangger, November 26, 1915. *CPAE*, Vol. 8, Doc. 152

You have no idea of the natural craftiness of such a woman. I would have been physically and mentally broken if I had not finally found the strength to keep her at arm's length and out of sight and earshot.

To Michele Besso, July 14, 1916. *CPAE*, Vol. 8, Doc. 233

She leads a worry-free life, has her two precious boys with her, lives in a fabulous neighborhood, does what she likes with her time, and innocently stands by as the guiltless party.

To Michele Besso, July 21, 1916. *CPAE*, Vol. 8, Doc. 238

The only thing she is missing is someone to dominate her. . . . What man would tolerate something so palpably smelly being stuck up his nose all his life, for no purpose at all, with the secondary obligation of also putting on a friendly face?

Ibid.

From now on I will no longer bother her about a divorce. The accompanying battle with my relatives has taken place. I have learned to withstand the tears.

> To Michele Besso, September 6, 1916. Einstein's relatives did not approve of his leaving his marriage in limbo, feeling it would compromise young Ilse's (Elsa's elder daughter's) eligibility for marriage. The divorce finally did take place in February 1919 in Switzerland. Einstein, as the guilty party, was ordered not to marry for the next two years; but, despite the ban, he married Elsa just two and a half months later since the prohibition did not apply under German law. *CPAE*, Vol. 8, Doc. 254; Fölsing, *Albert Einstein*, 425, 427

Separation from Mileva was a matter of life and death for me. . . . Thus I deprive myself of my boys, whom I still love tenderly.

> To Helene Savić, September 8, 1916. *CPAE*, Vol. 8, Doc. 258

I believe that Mitsa [Mileva] sometimes suffers from too great a reserve. Her parents and her sister . . . did not even know her address. In this respect, dear Helene, you could be of great use to her, helping her surmount her moments of discouragement. I am deeply grateful for everything you have done for Mitsa and especially for the children.

> Ibid.

I've been so preoccupied with what would happen in the event of my death that I'm surprised to find myself still alive.

To Mileva, April 23, 1918, after attending to legal paper-
work that would financially take care of her and the boys
in case of his death. *CPAE*, Vol. 8, Doc. 515

Mileva was absolutely insufferable when we were
together. When we are not, I can like her quite well;
she seems all right to me, even as the mother of my
boys.

To Michele Besso, July 29, 1918. *CPAE*, Vol. 8, Doc. 591

*From here it is difficult for me to determine whether
it might be better for Mileva and the boy [Eduard] to
move to her former homeland of Yugoslavia, . . .
where she and the boy could have an easier existence than in expensive Switzerland. . . . I cannot
give them any more help because the political situation has put all my relatives and my circle of friends
into extreme hardship as well, so that I've reached
my limits.

To Heinrich Zangger, September 18, 1938. Einstein Archives 40-116

She never reconciled herself to the separation and
divorce, and a disposition developed reminiscent of
the classical example of Medea. This darkened the
relations with my two boys, to whom I was attached
with tenderness. This tragic aspect of my life continued undiminished until my advanced age.

To Carl Seelig, May 5, 1952, about Mileva. Einstein Archives 39-020

Einstein began a long-distance affair with his cousin Elsa, who lived in Berlin, in 1912, while he was still married to Mileva and living in Zurich. The affair continued after the family moved to Berlin in 1914. He was not divorced from Mileva, who soon returned to Zurich, until February 1919. In June of that year he married Elsa, though for many years he had been telling friends that he had no intention of marrying her and had even considered marrying her daughter Ilse instead. At one time he had also had his eye on Paula, Elsa's younger sister. See various letters in *CPAE*, Vol. 8, and Stern, *Einstein's German World*, 105n.

I will always destroy your letters, as is your wish. I have already destroyed the first one.

To Elsa, April 30, 1912, responding to her misgivings about their affair. *CPAE*, Vol. 5, Doc. 389

I must love someone. Otherwise it is a miserable existence. And that someone is you.

Ibid.

I suffer even more than you because you suffer only for what you do not have.

To Elsa, May 7, 1912, alluding to his difficult wife, Mileva. *CPAE*, Vol. 5, Doc. 391

I am writing so late because I have misgivings about our affair. I have a feeling that it will not be good for us, nor for the others, if we form a closer attachment.

To Elsa, May 21, 1912. *CPAE*, Vol. 5, Doc. 399

I now have someone about whom I can think with unrestrained pleasure and for whom I can live. . . . We will have each other, something we have missed so terribly, and will give each other the gift of stability and an optimistic view of the world.

To Elsa, October 10, 1913. *CPAE*, Vol. 5, Doc. 476

If you were to recite for me the most beautiful poem . . . my pleasure would not even approach the pleasure I felt when I received the mushrooms and goose cracklings you prepared for me; . . . you will surely not despise the domestic side of me that is revealed by this disclosure.

To Elsa, November 7, 1913. *CPAE*, Vol. 5, Doc. 482

I really delight in my local relatives, especially in a cousin of my age, with whom I am linked by an old friendship. It is mostly because of this that I am accustoming myself very well to the large city [Berlin], which is otherwise loathsome to me.

To Paul Ehrenfest, ca. April 10, 1914, on his adjustment to life in Berlin. *CPAE*, Vol. 8, Doc. 2

*She [Elsa] was the main reason I came to Berlin.

> To Heinrich Zangger, June 27, 1915. *CPAE*, Vol. 8, Doc. 16a, embedded in Vol. 10

*I decided on the formality of marriage with my cousin [Elsa] after all, because her adult daughters would otherwise be seriously harmed. It doesn't signify any injury either to me or my boys, but is my duty. . . . Nothing in my life changes by it. Why should the original sin be even harder on these poor daughters of Eve?

> To Heinrich Zangger, March 1, 1916. The Einsteins felt that Margot and Ilse would have difficulty finding marriage partners if their mother was carrying on an affair. Yet in a letter to Michele Besso on December 5, he once again claimed, "I abandoned once and for all the idea of remarrying." The marriage in fact took place three years later. *CPAE*, Vol. 8, Docs. 196a, and 283a, respectively, both embedded in Vol. 10

I would take only one of the women with me, either Elsa or Ilse. The latter is more suitable because she is healthier and more practical.

> To Fritz Haber, October 6, 1920, on taking a traveling companion on a lecture trip to Norway. Einstein Archives 12-325. He neglected to mention that he had also been infatuated with Ilse, who was Elsa's daughter, before Elsa's and his marriage (see Ilse's letter to Georg Nicolai, May 22, 1918, *CPAE*, Vol. 8, Doc. 545).

With Mileva, Einstein had two sons, Hans Albert and Eduard, and a daughter, referred to as "Lieserl"; by his marriage to Elsa, he had two stepdaughters, Ilse and Margot. Lieserl was born in January 1902 before Einstein and Mileva were married, and she may have been given up for adoption or was raised by friends. She may also have died from the effects of scarlet fever when she was a toddler; no mention is made of her after September 1903, and Einstein apparently never saw her. See *CPAE*, Vol. 5, and *The Love Letters*. There is still speculation as to the real fate of Lieserl, who may have survived the scarlet fever and never knew her origins, or suffered some other yet unknown fate. Only Hans Albert had children. Eduard developed schizophrenia at the age of twenty, though up to that time he had been a somewhat fragile but essentially healthy young man pursuing a medical education. Eduard remained in Switzerland all of his life; Einstein told his biographer Carl Seelig that he rarely wrote to Eduard after leaving Europe for reasons he could not analyze himself. Einstein Archives 39-060. See also the section "On and to Children" for letters to his sons when they were children.

I'm very sorry about what has befallen Lieserl. It's so easy to suffer lasting effects from scarlet fever. If this will only pass. As what is the child registered? We must take precautions that problems don't arise for her later.

This somewhat cryptic (to the reader) letter was sent to Mileva ca. September 19, 1903. Registering a child may indicate the parents' intention of giving it up for adoption. They may have considered Lieserl's illegitimacy a threat to Einstein's provisional federal appointment at the Swiss Patent Office. See *CPAE*, Vol. 5, Doc. 13, n. 4

At the time we [he and Mileva] were separating from each other, the thought of leaving the children stabbed me like a dagger every morning when I awoke, but I have never regretted the step in spite of it.

To Heinrich Zangger, November 26, 1915. *CPAE*, Vol. 8, Doc. 152

Albert is now gradually entering the age at which I can mean very much to him. . . . My influence will be limited to the intellectual and esthetic. I want to teach him mainly to think, judge, and appreciate things objectively. For this I need several weeks a year—a few days would only be a short thrill with no deeper value.

To Mileva, who was afraid that her own relationship with Hans Albert would suffer if he had too much contact with his father, December 1, 1915. *CPAE*, Vol. 8, Doc. 159

My compliments on the good condition of our boys. They are in such excellent physical and emotional shape that I could not wish for more. And I know this is for the most part due to the proper upbringing you are providing. . . . They came to meet me spontaneously and sweetly.

To Mileva during his visit to Zurich, April 8, 1916. *CPAE*, Vol. 8, Doc. 211

*My relations with the boys have frozen up completely again. Following an exceedingly nice Easter excursion, the subsequent days in Zurich brought on a complete chilling in a way that is not quite explicable to me. It's better if I keep my distance from them; I have to content myself with the knowledge that they are developing well. How much better off I am than countless others who have lost their children in the War!

To Heinrich Zangger, July 11, 1916. Zangger was a close friend of Einstein's who kept an eye on his boys. Hans Albert lived with him occasionally when Mileva was ill. *CPAE*, Vol. 8, Doc. 232, embedded in Vol. 10

Is [Hans] Albert with you yet? I miss him often. He is already a person with a mind of his own whom one can talk to, and so thoroughly sound in an honest way. He rarely writes but I understand it's not his sort of thing. . . . It's good he did not grow up in the big city with its superficiality.

To Heinrich Zangger, December 24, 1919. *CPAE*, Vol. 9, Doc. 233

Maybe I can muster enough foreign money to be able to let them [Mileva and the children] stay in Zurich. This may have advantages for my children's more distant future, which would justify tackling the difficulties.

> To Michele Besso, January 6, 1920. Because of the unfavorable Swiss/German exchange rate, Einstein had been contemplating asking them to move to southern Germany, where his money would go farther. *CPAE*, Vol. 9, Doc. 245

I still don't know when I can come to Switzerland. . . . I am delighted that Albert is with you. I am going to get European money again soon for my family; nothing more can be done with the local currency. . . . Albert should not get the feeling that his Papa doesn't worry about his upkeep.

> To Heinrich Zangger, February 27, 1920, alluding to the deflation of the German mark. *CPAE*, Vol. 9, Doc. 332

*My otherwise so cheerful boys are somewhat painfully envious of both of you. . . . They feel as if I have exchanged them for you.

> To Margot Einstein, referring to her and Ilse, August 26, 1921. *CPAE*, Vol. 12, Doc. 214

*I feel the need to thank you for the fine days I was allowed to spend with our dear boys. I'm grateful that you raised them in a friendly frame of mind toward me and in an exemplary manner otherwise as well. I am most especially satisfied with their cheerful and modest ways; secondly, of course, also their lively intelligence.

> To Mileva, August 28, 1921. *CPAE*, Vol. 12, Doc. 218

*[Eduard's] intellectual abilities may be even stronger [than Hans Albert's], but he seems to lack equilibrium and a sense of responsibility (too egoistic). Too little personal interaction with others, which gives rise to a feeling of isolation and inhibitions of other sorts. He is an interesting little fellow, but he will not have an easy time in life.

> To Mileva, August 14, 1925. Einstein Archives 75-963

I could be a grandpa now too if my [Hans] Albert hadn't married such a *Schlemilde*.

> To his uncle, Caesar Koch, who had just become a grandfather, October 26, 1929. Einstein Archives 47-271. Einstein had vigorously opposed Hans Albert's marriage to Frieda Knecht, who was nine years his senior; but the couple remained together until Frieda's death. By 1930 they had made Einstein a grandpa, too, with the birth of Bernhard. See Sotheby's auction catalog, June 26, 1998, 424

*My defiant attitude prevented me from noticing how much I have suffered from the personal quarrels with all of you.

> To Mileva, June 15, 1933. Translated in Neffe, *Einstein*, 199. Einstein Archives 75-962

The deepest sorrow loving parents can experience has come upon you. Everything I observed of your little son indicated he was becoming a well-balanced and self-assured person with a healthy outlook on life. Although I saw him only for a short period, he was as close to me as if he had grown up near me.

> To Hans Albert and wife Frieda, January 7, 1939, after the sudden death of six-year-old Klaus, their son, probably of diphtheria. See Roboz Einstein, *Hans Albert Einstein*, 34

It is a thousand pities for the boy that he must pass his life without the hope of a normal existence. Since the insulin injections have proved unsuccessful, I have no further hopes from the medical side. . . . I think it is better on the whole to let Nature run its course.

> To Michele Besso, November 11, 1940, about son Eduard. Einstein Archives 7-378

There is a block behind it that I cannot fully analyze. But one factor is that I think I would arouse painful feelings of various kinds in him if I made an appearance in whatever form.

> To Carl Seelig, January 4, 1954, stating why he was not in touch with Eduard. In his will, Einstein left a larger amount of money to Eduard than to Hans Albert. Einstein Archives 39-059

It is a joy for me to have a son who has inherited the chief trait of my personality: the ability to rise above mere existence by sacrificing oneself through the years for an impersonal goal. This is the best, indeed the only way in which we can make ourselves independent from personal fate and from other human beings.

To Hans Albert, May 11, 1954. Einstein Archives 75-918

Honesty compels me to admit that Frieda reminded me of your 50th birthday.

Ibid.

When Margot speaks, you see flowers growing.

Commenting on his stepdaughter Margot's love of nature. Quoted by friend Frieda Bucky in "You Have to Ask Forgiveness," *Jewish Quarterly* 15, no. 4 (Winter 1967–68), 33

ABOUT HIS SISTER, MAJA,
AND MOTHER, PAULINE

Yes, but where are its wheels?

Two-and-a-half-year-old Albert, after the birth of Maja in 1881, upon being told he would now have something new to play with. In "Biographical Sketch," by Maja Winteler-Einstein, *CPAE*, Vol. 1, lvii

My mother and sister seem somewhat petty and philistine to me, despite the sympathy I feel for them. It is interesting how life gradually changes us in the very subtleties of our soul, so that even the closest of family ties dwindle into habitual friendship. Deep inside we no longer understand one another and are incapable of empathizing with the other, or know what emotions move the other.

To Mileva Marić, early August 1899, at age twenty. *The Love Letters*, 9; *CPAE*, Vol. 1, Doc. 50

My poor mother arrived here on Sunday. . . . Now she is lying in my study and suffering terribly, physically and mentally. . . . It seems that her torments will last a long time yet; she still looks good, but mentally she has suffered very much under the morphine.

To Heinrich Zangger, January 3, 1920. *CPAE*, Vol. 9, Doc. 242

My mother died a week ago today in terrible agony. We are all completely exhausted. One feels in one's bones the significance of blood ties.

To Heinrich Zangger, February 27, 1920. *CPAE*, Vol. 9, Doc. 332; Einstein Archives 39-732

I know what it means to see one's mother suffer the agony of death and not be able to help. There is no consolation. All of us have this heavy burden to bear, for it is inseparably bound up with life.

To Hedwig Born, June 18, 1920, after the death of her mother. In Born, *Born-Einstein Letters*, 28. *CPAE*, Vol. 10, Doc. 59

On Aging

I have remained a simple fellow who asks nothing of the world; only my youth is gone—the enchanting youth that forever walks on air.

To Anna Meyer-Schmid, May 12, 1909. *CPAE*, Vol. 5, Doc. 154

I lived in that solitude which is painful in youth, but delicious in maturity.

From a statement written for Georges Schreiber's *Portraits and Self-Portraits* (Boston: Houghton Mifflin, 1936). Reprinted in *Out of My Later Years*, 13. Einstein Archives 28-332

There is, after all, something eternal that lies beyond the reach of the hand of fate and of all human delusions. And such eternals lie closer to an older person than to a younger one who oscillates between fear and hope.

To Queen Elisabeth of Belgium, March 20, 1936. Einstein Archives 32-387

*At our age, the devil doesn't give you much time off!

To Heinrich Zangger, February 27, 1938. Quoted in Seelig, *Helle Zeit, Dunkle Zeit*, 45, and translated in Neffe, *Einstein*, 199. Einstein Archives 40-105

People like you and I, though mortal of course, like everyone else, do not grow old no matter how long we live. What I mean is that we never cease to stand like curious children before the great Mystery into which we were born.

> To Otto Juliusburger, September 29, 1942. Einstein Archives 38-238

*Your "I feel too old" I am not taking too seriously, because I know this feeling myself. Sometimes . . . it surges upwards and then subsides again. We can after all quietly leave it to nature gradually to reduce us to dust if she does not prefer a more rapid method.

> To Max Born, September 7, 1944. In Born, *Born-Einstein Letters*, 145. Einstein Archives 8-207

*Though I am now an old fogey, I am still hard at work and still refuse to believe that God plays dice.

> To former Berlin student Ilse Rosenthal-Schneider, May 11, 1945. Einstein Archives 20-274

I am content in my later years. I have kept my good humor and take neither myself nor the next person seriously.

> To P. Moos, March 30, 1950. Einstein Archives 60-587

All of one's contemporaries and aging friends are living in a delicate balance, and one feels that one's own consciousness is no longer as brightly lit as it once was. But then, twilight with its more subdued colors has its charms as well.

To Gertrud Warschauer, April 4, 1952. Einstein Archives 39-515

I [have] always loved solitude, a trait that tends to increase with age.

To E. Marangoni, October 1, 1952. Einstein Archives 60-406

If younger people were not taking care of me, I would surely try to be institutionalized, so that I would not have to become so concerned about the decline of my physical and mental powers, which after all is unpreventable in the natural course of things.

To W. Lebach, May 12, 1953. Einstein Archives 60-221

I feel like an egg, of which only the shell remains— at 75 years old, one can't expect anything else. One should prepare a person for his death.

Quoted by Fantova, "Conversations with Einstein," January 1, 1954

In one's youth every person and every event appear to be unique. With age, one becomes much more aware that similar events recur. Later on, one is less often delighted or surprised, but also less disappointed.

> To Queen Elisabeth of Belgium, January 3, 1954. Einstein Archives 32-408

I believe that older people who have scarcely any-thing to lose ought to be willing to speak out on be-half of those who are young and who are subject to much greater restraint.

> To Queen Elisabeth of Belgium, March 28, 1954. Einstein Archives 32-411

I am feeling my age greatly. I'm no longer so eager to work and always have to lie down after a meal. I enjoy living, but I would not mind if it all suddenly ended.

> Quoted by Fantova, "Conversations with Einstein," April 27, 1954

I no longer have the strong pains I had earlier, but I feel very weakened, as can be expected of such an old geezer.

> Ibid., May 29, 1954

Today [due to illness] I stayed in bed and received guests like an old lady of the eighteenth century. This was fashionable in Paris at that time. But I'm not a woman, and this isn't the eighteenth century!

> Ibid., June 11, 1954

I'm like a run-down old car—something is wrong in every corner. But life is still worthwhile as long as I can still work.

Ibid., January 9, 1955

Even [old] age has very beautiful moments.

To Margot Einstein. Quoted in Sayen, *Einstein in America*, 298

On America and Americans

I am happy to be in Boston. I have heard of Boston as one of the most famous cities in the world and the center of education. I am happy to be here and expect to enjoy my visit to this city and to Harvard.

On his visit to the city with Chaim Weizmann. *New York Times*, May 17, 1921. Contributed by A. J. Kox in response to the many quotations about Princeton in this book (see later in this section).

*America is interesting, with all its hustle and bustle. It is easier to feel enthusiasm for it than for other countries I've unsettled with my presence. I had to consent to being shown around like a prize ox to address innumerable small and large gatherings. . . . It's a wonder I survived it all.

To Michele Besso, ca. May 21–30, 1921. *CPAE*, Vol. 12, Doc. 141

*It is the women . . . who dominate all of American life. The men are interested in nothing at all; they work, work as I haven't seen anyone work anywhere else. For the rest, they are toy dogs for their wives, who spend the money in the most excessive fashion and who shroud themselves in a veil of extravagance.

From an interview in the *Nieuwe Rotterdamsche Courant*, July 4, 1921. Einstein insisted he was wrongly quoted and wrote a rebuttal in the *Vossische Zeitung* six days later, claiming he was shocked when he read the account.

See Rowe and Schulmann, *Einstein on Politics,* 111–112, for an account of the fiasco.

Even if Americans are less scholarly than Germans, they do have more enthusiasm and energy, causing a wider dissemination of new ideas among the people.

Quoted in the *New York Times,* July 12, 1921

A firm approach is indispensable everywhere in America; otherwise one receives no payment and little esteem.

To Maurice Solovine, January 14, 1922. Published in *Letters to Solovine,* 49. Einstein Archives 21-157

Never yet have I experienced from the fair sex such energetic rejection of my advances; or if I have, never from so many at once.

Part of a reply to the Women Patriot Corporation, via the Associated Press, December 1932. According to Rowe and Schulmann, *Einstein on Politics,* 261–262, the right-wing group, under the leadership of a Mrs. Frothingham, had protested Einstein's American visit to the State Department, arguing that he was the ringleader of an anarcho-communist plot, and that the theory of relativity was subversive and designed to promote lawlessness and shatter Church and State. Einstein Archives 28-213

*But are they not quite right, these watchful citizenesses? Why should one open one's doors to a person who devours hardboiled capitalists with as much appetite and gusto as the Cretan Minotaur in days gone by devoured luscious Greek maidens, and on top of that is low-down enough to reject every sort of war, except the unavoidable war with one's own wife?

Ibid., 262

In America, more than anywhere else, the individual is lost in the achievements of the many.

From an interview with G. S. Viereck, "What Life Means to Einstein," *Saturday Evening Post*, October 26, 1929; reprinted in Viereck, *Glimpses of the Great*, 438

Americans undoubtedly owe much to the Melting Pot. It is possible that this mixture of races makes their nationalism less objectionable than the nationalism of Europe. . . . It may be due to the fact that [Americans] do not suffer from the heritage of hatred or fear, which poisons the relations of the nations of Europe.

Ibid., 451

*Here, everyone stands up proudly and jealously for his civil rights. Everyone, irrespective of birth, has the opportunity, not merely on paper but in actual practice, to develop his energies freely for the benefit of the human community as a whole. . . . Individual freedom provides a better basis for productive labor than any form of tyranny.

From a shipside broadcast to the American people on arrival in New York, December 11, 1930. See Rowe and
Schulmann, *Einstein on Politics*, 238–239, for full text. Einstein Archives 36-306

*I salute you and the soil of your country. I eagerly
look forward to renewing old friendships and to
broadening my understanding in the light of what I
shall see and learn while among you.

Ibid.

I feel that you are justified in looking into the future
with true assurance, because you have a mode of
living in which one finds the joy of life and the joy of
work harmoniously combined. Added to this is the
spirit of ambition which pervades your very being,
and seems to make the day's work like a happy
child at play.

From a New Year's Day greeting, in the *New York Times*,
January 1, 1931. Quoted in *Stevenson's Book of Quotations:
Classical and Modern*. Einstein Readex 324 (not in archive
database)

Here in Pasadena it is like paradise. . . . Always sun-
shine and fresh air, gardens with palm and pepper
trees, and friendly people who smile at one and ask
for autographs.

To the Lebach family during the days before smog, January 16, 1931, on the city in which the California Institute of
Technology is located. Einstein Archives 47-373

[America], this land of contrasts and surprises, which leaves one filled alternately with admiration and incredulity. One feels more attached to the Old Europe, with its heartaches and hardships, and is glad to return there.

To Queen Elisabeth of Belgium, February 9, 1931, revealing a touch of homesickness during his three-month stay in America. Einstein Archives 32-349

The smile on the faces of the people . . . is symbolic of one of the greatest assets of the American. He is friendly, self-confident, optimistic—and not envious.

From "Impressions of the U.S.A," ca. 1931. Source misquoted in *Ideas and Opinions*, 3. Einstein Archives 28-167

The American lives even more for his goals, for the future, than the European. Life for him is always becoming, never being. . . . He is less of an individualist than the European . . . more emphasis is put on the "we" than the "I."

Ibid.

I have warm admiration for American institutes of scientific research. We are unjust in attempting to ascribe the increasing superiority of American research work exclusively to superior wealth; devotion, patience, a spirit of comradeship, and a talent for cooperation play an important part in its success.

Ibid.

This proves that knowledge and justice are ranked above power and wealth by a large section of the human race.

> Ibid. Einstein came to this conclusion because Americans showed such reverence and respect for him, despite their reputed materialism.

For the long term I would prefer being in Holland rather than in America. . . . Besides having a handful of really fine scholars, it is a boring and barren society that would soon make you tremble.

> To Paul Ehrenfest, April 3, 1932, after his return to Europe. Einstein Archives 10-227

I am very happy at the prospect of becoming an American citizen in another year. My desire to be a citizen of a free republic has always been strong and prompted me in my younger days to emigrate from Germany to Switzerland.

> From a statement issued on his sixtieth birthday. *Science* 89, n.s. (1939), 242

*In America, the development of the individual and his creative powers is possible, and that, to me, is the most valuable asset in life.

> From "I Am an American," June 22, 1940. See Rowe and Schulmann, *Einstein on Politics*, 470. Einstein Archives 29-092

*From what I have seen of Americans since I came here, . . . they are not suited, either by temperament or tradition, to live under a totalitarian system. I believe that many of them would find life not worth living under such circumstances. Hence, it is all the more important for them to see to it that these liberties be preserved and protected.

Ibid.

*I believe that America will prove that democracy is not merely a form of government based on a sound Constitution but is, in fact, a way of life tied to a great tradition, the tradition of moral strength. Today more than ever, the fate of the human race depends upon the moral strength of human beings.

Ibid., 472

America is today the hope of all honorable men who respect the rights of their fellow men and who believe in the principles of freedom and justice.

From "Message for Germany," dictated over the telephone on December 7, 1941, the day Pearl Harbor was bombed, to a White House correspondent. Quoted in Nathan and Norden, *Einstein on Peace*, 320. Einstein Archives 55-128

The only justifiable purpose of political institutions is to assure the unhindered development of the individual. . . . That is why I consider myself to be particularly fortunate to be an American.

Ibid.

[The United States has] a government controlled to a large extent by financiers, the mentality of whom is near to the fascist frame of mind. If Hitler were not a lunatic, he could easily have avoided the hostility of the Western powers.

To Frank Kingdon, September 3, 1942. Einstein Archives 55-469

*In the United States everyone feels assured of his worth as an individual. No one humbles himself before another person or class. Even the great difference in wealth, the superior power of a few cannot undermine this healthy self-confidence and natural respect for the dignity of one's fellow-man.

From "Message to My Adopted Country," *Pageant* 1, no. 12 (January 1946), 36–37. See Rowe and Schulmann, *Einstein on Politics*, 474; Jerome and Taylor, *Einstein on Race and Racism*, 140

The separation [between Jews and Gentiles] is even more pronounced [in America] than it ever was anywhere in Western Europe, including Germany.

To Hans Muehsam, March 24, 1948. Einstein Archives 38-371

I hardly ever felt as alienated from people as I do right now. . . . The worst is that nowhere is there anything with which one can identify. Brutality and lies are everywhere.

> To Gertrud Warschauer, July 15, 1950, about the McCarthy era. Einstein Archives 39-505

The German calamity of years ago repeats itself: people acquiesce without resistance and align themselves with the forces of evil.

> To Queen Elisabeth of Belgium, January 6, 1951, about Mc-Carthyism in America. Einstein Archives 32-400

I have become a kind of enfant terrible in my new homeland because of my inability to keep silent and swallow everything that happens here.

> To Queen Elisabeth of Belgium, March 28, 1954. Einstein Archives 32-410

ON HIS ADOPTED HOMETOWN OF PRINCETON,
NEW JERSEY

I found Princeton lovely: an as yet unsmoked pipe, so fresh, so young.

> Quoted in the *New York Times*, July 8, 1921, reporting on his lecture trip to his future hometown

Princeton is a wondrous little spot, a quaint and cer-
emonious village of puny demigods on stilts. Yet, by
ignoring certain social conventions, I have been able
to create for myself an atmosphere conducive to
study and free from distraction.

> To Queen Elisabeth of Belgium, November 20, 1933. Ein-
> stein Archives 32-369

To an elderly man society here remains intrinsically
foreign.

> To Queen Elisabeth of Belgium, February 16, 1935. Einstein
> Archives 32-385

I am very happy with my new home in friendly
America and in the liberal atmosphere of Princeton.

> From an interview, *Survey Graphic* 24 (August 1935), 384,
> 413

I am privileged by fate to live here in Princeton as if
on an island that . . . resembles the charming palace
garden in Laeken [Belgium]. Into this small university town the chaotic voices of human strife barely
penetrate. I am almost ashamed to be living in such
a place while all the rest struggle and suffer.

> To Queen Elisabeth of Belgium, March 20, 1936. Einstein
> Archives 32-387

In the face of all the heavy burdens I have borne in
recent years, I feel doubly thankful that there has

fallen on my lot in Princeton University a place for work and a scientific atmosphere which could not be better or more harmonious.

> To university president Harold Dodds, January 14, 1937. At the time, Einstein's office was temporarily located on the Princeton campus even though he was a member of the Institute for Advanced Study, a separate institution whose campus had not yet been built. Part of this message is inscribed on the Einstein statue in Princeton. Einstein Archives 52-823

The Marquand estate is now a public park, and because today was Sunday and I didn't go to the Institute, I took a walk there—it's so close by, and so beautiful.

> Quoted by Fantova, "Conversations with Einstein," May 8, 1954

A banishment to paradise.

> On going to Princeton. Quoted in Sayen, *Einstein in America*, 64

You are surprised, aren't you, at the contrast between my fame throughout the world . . . and the isolation and quiet in which I live here. I wished for this isolation all my life, and now I have finally achieved it here in Princeton.

> Quoted in Frank, *Einstein: His Life and Times*, 297

On and to Children

Let me tell you what I look like: pale face, long hair, and a tiny start of a paunch. In addition, an awkward gait, and a cigar in the mouth . . . and a pen in pocket or hand. But crooked legs and warts he does not have, and so is quite handsome—also there's no hair on his hands, as is so often the case with ugly men. So it really is a pity that you didn't see me.

> From a postcard to eight-year-old cousin Elisabeth Ney, who felt neglected because she wasn't invited for a visit with Einstein along with her parents, September 30, 1920. In Calaprice, *Dear Professor Einstein*, 113. *CPAE*, Vol. 10, Doc. 157

*Dear young man: In your article you were quite right in stating that motion can be experienced and presented by us only as relative motion. . . . But until the presentation of relativity theory, it was thought that the concept of absolute motion was necessary for the formulation of the laws of motion. . . . But it would be better if you began to teach others only after you yourself have learned something.

> To Arthur Cohen, age twelve, who had submitted a paper to Einstein, December 26, 1928. Einstein Archives 42-547. Cohen's sister-in-law, Betty, contacted me after reading this quotation. Young Arthur eventually went to Stanford, then received a Ph.D. in botany from Harvard, and became a professor at Washington State University.

*Dear little cousin: You are not the most savvy little customer, but it's good that you're at least a curious young fellow. So then: The soup doesn't cool down as much because the layer of fat on the top makes evaporation more difficult and thereby also slows the cooling.

> To an inquisitive child, name unknown, January 13, 1930. See Calaprice, *Dear Professor Einstein*, 121. Einstein Archives 42-592

*To the schoolchildren of Japan: . . . I myself have visited your beautiful country, its cities, houses, its mountains and forests, from which Japanese youngsters derive a love for their homeland. On my table lies a large book full of colorful pictures drawn by Japanese children. . . . Remember that ours is the first era in which it has been possible for people of different nations to conduct their affairs in a friendly and understanding manner. . . . May the spirit of brotherly understanding . . . continue to grow. . . . I, an old man, . . . hope that your generation will some day put mine to shame.

> Written in fall 1930. In Calaprice, *Dear Professor Einstein*, 122–123. Einstein Archives 42-594

*My dear young people: Bear in mind the fact that the wonderful things which you come to know in your schools are the product of many generations . . . accomplished in enthusiastic struggle and with great effort in all countries of the earth. All this is now laid in your hands as your inheritance, to the end that you may receive, honor, and advance it and some day faithfully convey it to your posterity . . . If you will constantly keep this in mind, you will find a meaning in life and effort and will attain the right attitude toward other peoples and other times.

> From a speech at Pasadena City College, February 26, 1931. As published in *Pasadena Star News*, February 26, 1931, and *Pasadena Chronicle*, February 27, 1931. Differently translated in *Mein Weltbild*, 25. See also further information about this speech in the section "On Education, Students, and Academic Freedom."

*Despite the natural presumption that life in some form may not be unique to our planet, the notion is beyond the current realm of knowledge.

> To Dick Emmons, a sixteen-year-old amateur astronomer who later became a longtime member of Operation Moonwatch, a Smithsonian-sponsored organization started in 1956 to help track artificial satellites, November 11, 1935. Smithsonian Institution Archives, Richard Emmons Collection, Record Unit 08-112, Box 1. Also see Patrick Mc-Cray, *Keep Watching the Skies!* (Princeton, N.J.: Princeton University Press, 2008), 35. Einstein Archives 92-381

*Dear Children: With pleasure I can picture you children gathered all together during the holidays, united by a harmonious spirit instilled by the warm glow of Christmas lights. But also remember the lessons taught by the one whose birthday you are celebrating. . . . Learn to be happy through the good fortunes and joys of your friends and not through senseless quarrels. . . . Your burden will seem lighter or more bearable to you, you will find your own way through patience, and you will spread joy everywhere.

> In response to a school's request for a meaningful Christmas message, December 20, 1935. In Calaprice, *Dear Professor Einstein*, 134–135. Einstein Archives 42-598

*Dear Phyllis: . . . . Scientists believe that every occurrence, including the affairs of human beings, is due to the laws of nature. Therefore a scientist cannot be inclined to believe that the course of events can be influenced by prayer. . . . But also, everyone who is seriously involved in the pursuit of science becomes convinced that some spirit is manifest in the laws of the universe, one that is vastly superior to that of man. In this way the pursuit of science leads to a religious feeling of a special sort.

> To Phyllis Wright, January 24, 1936. In Calaprice, *Dear Professor Einstein*, 128–129. Einstein Archives 42-602

*Dear Barbara: I was very pleased with your kind letter. Until now I never dreamed to be something like a hero. But since you have given me the nomination, I feel that I am one. . . . Do not worry about your difficulties in mathematics; I can assure you that mine are still greater.

> To junior high school student Barbara Wilson, January 7, 1943. In Calaprice, *Dear Professor Einstein*, 140. Einstein Archives 42-606

*Dear Hugh: There is no such thing like an irresistible force and immovable body. But there seems to be a very stubborn boy who has forced his way victoriously through strange difficulties created by himself for this purpose.

> To Hugh Everett III, June 11, 1943. Hugh's letter is not in
> the archive, so there is no further background information.
> Einstein Archives 89-878

*Dear Myfanwy: . . . I have to apologize to you that I am still among the living. There *will* be a remedy for this, however. Be not worried about "curved space." You will understand [it] at a later time. . . . Used in the right sense the word "curved" has not exactly the same meaning as in everyday language. . . . I hope that [your] future atsronomical investigations will not be discovered anymore by the eyes and ears of your school government.

> To Myfanwy Williams in South Africa, who confided that
> she thought Einstein was dead, and that she and her
> friends secretly used a telescope after lights-out at her
> boarding school, August 25, 1946. Her first name had been
> wrongly transcribed as "Tyfanny"—thanks to Barbara
> Wolff of the Einstein Archives for the correction. In Cala-
> price, *Dear Professor Einstein*, 153. Einstein Archives 42-612

*I do not mind that you are a girl. But the main thing is that you yourself do not mind. There is no reason for it.

> From another letter to Myfanwy, September–October 1946.
> Einstein had mistaken her name for a boy's, and she wrote

back explaining she's a girl, something she had "always regretted" but had become "resigned to." In Calaprice, *Dear Professor Einstein*, 156. Einstein Archives 42-614

*Dear Monique: There has been an earth since a little more than a billion years. As for the question of the end of it I advise: Wait and see! . . . I enclose a few stamps for your collection.

To Monique Epstein in New York, June 19, 1951. In Calaprice, *Dear Professor Einstein*, 174–175. Einstein Archives 42-647

*Dear Children: . . . Without sunlight there is: no wheat, no bread, no grass, no cattle, no meat, no milk, and everything would be frozen. No LIFE.

To "Six Little Scientists" in Louisiana, December 12, 1951. In Calaprice, *Dear Professor Einstein*, 187. Einstein Archives 42-652

*As an old schoolmaster I received with great joy and pride the nomination to the office of Rectorship of your Society. Despite my being an old gypsy there is a tendency to respectability inherent in old age— also with me. . . . I am a little (but not too much) bewildered . . . that this nomination was made independent of my consent.

To Sixth Form Society, Newcastle-under-Tyne, March 17, 1952. Einstein Archives 42-660

*You young people should consider yourselves fortunate that you . . . have the opportunity to exchange viewpoints and ideas with those of a variety of cultural backgrounds. There is no better opportunity to acquire the lifelong insights that are necessary for the resolution of international problems and conflicts.

From a message to Friendship among Children and Youth in Austria, November 22, 1952. In Calaprice, *Dear Professor Einstein*, 203. Einstein Archives 42-667.1

*Dear Children: . . . We call something an animal which has certain characteristics: it takes nourishment, it descends from parents similar to itself, it grows, it moves by itself, it dies if its time has run out. . . . Think about [humans] in the above-mentioned way and then decide for yourselves whether it is a natural thing to regard ourselves as animals.

To children of Westview School, January 17, 1953. In Calaprice, *Dear Professor Einstein*, 206. Einstein Archives 42-673

*Your gift will be an appropriate suggestion to be a little more elegant in the future . . . because neckties and cuffs exist for me only as remote memories.

To children of Farmingdale, N.Y., Elementary School, March 26, 1955, less than a month before his death. In Calaprice, *Dear Professor Einstein*, 219–220. Einstein Archives 42-711

Nowhere else is it as nice for boys as in Zurich. . . .
Boys aren't pestered too much with homework
there, nor with the need to be too well dressed and
well mannered.

> To son Hans Albert, after the boys returned to Zurich with
> their mother, January 25, 1915. *CPAE*, Vol. 8, Doc. 48

Today I'm sending off some toys for you and Tete.
Don't neglect your piano, my Adn; you don't know
how much pleasure you can give to others, as well
as to yourself, when you can play music nicely. . . .
Another thing, brush your teeth every day, and if a
tooth is not quite all right, go to the dentist immediately. I also do the same and am now very happy
that I have healthy teeth. This is very important, as
you will realize yourself later on.

> To Hans Albert ("Adn," earlier mistranscribed in the ar-
> chive as "Adu"; "Tete" is Eduard), ca. April 1915. *CPAE*,
> Vol. 8, Doc. 70. In a letter later that year he urges the two
> boys to take calcium chloride after every meal to promote
> strong tooth and bone development.

I'll try to be together with you for a month every
year so that you will have a father who is close to

you and can love you. You can also learn a lot of good things from me that no one else can offer you so easily. The things I've gained from so much strenuous work should be of value not only to strangers but especially to my own boys. In the last few days I completed one of the finest papers of my life. When you're older I'll tell you about it.

> To eleven-year-old Hans Albert, November 4, 1915, also referring to his paper on the general theory of relativity. *CPAE*, Vol. 8, Doc. 134

On the piano, play mainly the things that you enjoy, even if your teacher doesn't assign them to you. You learn the most from things that you enjoy doing so much that you don't even notice that the time is passing. Often I'm so engrossed in my work that I forget to eat lunch.

> Ibid.

I'm very glad that you enjoy the piano so much. I have one in my little apartment, too, and play it every day. I also play the violin a lot. Maybe you can practice something to accompany a violin, and then we can play at Easter when we're together.

> To Hans Albert, March 11, 1916. Einstein was not able to come at Christmas because of the difficulty of crossing borders during wartime, so he planned a trip at Easter. *CPAE*, Vol. 8, Doc. 199

You still make so many writing errors. You must take care in that regard: it makes a bad impression when words are misspelled.

To Hans Albert, March 16, 1916. *CPAE*, Vol. 8, Doc. 202

I'm writing you now for the third time without receiving a reply from you. Don't you remember your father anymore? Are we never going to see each other again?

To Hans Albert, September 26, 1916. Einstein learned that the boys had become angry with him. They reconciled and continued to write occasionally, while Einstein visited about once a year during wartime. *CPAE*, Vol. 8, Doc. 261; see also Doc. 258

Don't worry about your marks. Just make sure that you keep up with the work and that you don't have to repeat a year. It is not necessary to have good marks in everything.

To Hans Albert, October 13, 1916. *CPAE*, Vol. 8, Doc. 263

Although I'm over here, you do have a father who loves you more than anything else and who is constantly thinking of you and caring about you.

Ibid., regarding their separation

*It hurts me often, too, that I have so little of both of you, but I'm a very busy man and can't get away much from here. The two of us have been together so rarely that I hardly know you at all, even though I'm your father. I'm sure you have only a quite vague impression of me, too.

To Eduard Einstein, August 1, 1920. *CPAE*, Vol. 10, Doc. 96

*No father can be treated the way Albert's letter treated me. His letter is full of mistrust, lack of respect, and a nasty attitude toward me. I truly did not deserve that and will not put up with it.

To Eduard Einstein, July 15, 1923, after a continuing tiff with Hans Albert, who had faithfully sided with his mother during separation and divorce. Einstein Archives 75-627

*I dislike very much that my children should be taught something that is contrary to all scientific thinking.

Recalled by Philipp Frank in *Einstein: His Life and Times*, 280, about Einstein's children's religious education

On Death

I have firmly resolved to bite the dust, when my time comes, with a minimum of medical assistance, and up to then I will sin to my wicked heart's content.

To Elsa Einstein, August 11, 1913. *CPAE*, Vol. 5, Doc. 466

This life is not such that we ought to complain when it comes to an end for us or for a loved one; rather, we may look back in satisfaction when it has been bravely and honorably withstood.

To Ida Hurwitz, November 22, 1919. *CPAE*, Vol. 9, Doc. 172

The old who have died live on in the young ones. Don't you feel this now in your bereavement, when you look at your children?

To Hedwig Born, June 18, 1920, after the death of her mother. In Born, *Born-Einstein Letters*, 28–29 (date therein wrongly given as April 18). *CPAE*, Vol. 10, Doc. 59

Our death is not an end if we have lived on in our children and the younger generation. For they are us; our bodies are only wilted leaves on the tree of life.

To the widow of Dutch physicist Heike Kamerlingh Onnes, February 25, 1926. Einstein Archives 14-389

Death is a reality. . . . Life ends definitely when the subject, by his actions, no longer affects his environment. . . . He can no longer add an iota to the sum total of his experience.

> From an interview with G. S. Viereck, "What Life Means to Einstein," *Saturday Evening Post*, October 26, 1929; reprinted in Viereck, *Glimpses of the Great*, 444–445

Neither on my deathbed nor before will I ask myself such a question. Nature is not an engineer or a contractor, and I myself am a part of Nature.

> In answer to a question concerning what facts would determine if his life was a success or failure, November 12, 1930. Quoted in Dukas and Hoffmann, *Albert Einstein, the Human Side*, 92. Einstein Archives 45-751

One lives one's life under constant tension, until it is time to go for good.

> To his sister, Maja, August 31, 1935. Quoted in *Einstein: A Portrait*, 42. Einstein Archives 29-417

It should be of comfort to you that a sudden farewell to this best of all worlds is something that one must wish for a loved one above all, so that things don't happen as in Haydn's *Farewell* Symphony, where one instrument of the orchestra vanishes after another.

> To Boris Schwarz, on the death of his father, 1945. Quoted in Holton and Elkana, *Albert Einstein: Historical and Cultural Perspectives*, 416. Einstein Archives 79-678

Is there not a certain satisfaction in the fact tha'
natural limits are set to the life of the individual,
so that at its conclusion it may appear as a work
of art?

Quoted in "Paul Langevin," *La Pensée*, n.s., no 12 (May–
June 1947), 13–14. Einstein Archives 5-150

*The death of our closest relatives reopens old child-
hood wounds. . . . It is left to each of us to deal with
our share of these alone.

To Hans Albert Einstein, August 4, 1948. Einstein Archives
75-836

I feel unable to participate in your projected TV
broadcast "The Last Two Minutes." It seems to me
not so relevant how people are to spend the last two
minutes before their final deliverance.

In answer to a request that he participate in a television
program on how some famous people would spend the
last two minutes of their lives, August 26, 1950. Einstein
Archives 60-684

I myself should also be dead already, but I am still
here.

To E. Schaerer-Meyer, July 27, 1951. Einstein Archives
60-525

Brief is this existence, like a brief visit in a strange
house. The path to be pursued is poorly lit by a

flickering consciousness whose center is the limiting and separating "I." . . . When a group of individuals becomes a "we," a harmonious whole, they have reached as high as humans can reach.

> Obituary for physicist Rudolf Ladenburg, April 1952. See Stern, *Einstein's German World*, 163. Einstein Archives 5-160

*Man stands powerless before all the tragedies that are allotted to him. . . . But all the suffering unites us with those . . . with whom we so seldom have an opportunity to share a small part of our lives. . . . With my heart I touch your hand.

> To Gerhard Fankhauser, a professor of biology at Princeton, whose wife died in a car accident. Einstein also asked that Fankhauser's young children visit him at home. November 10, 1954. Sent to me by Robin Remy, whose mother inherited a copy of the original. Einstein Archives 59-630

To one bent on age, death will come as a release. I feel this quite strongly now that I have grown old myself and have come to regard death like an old debt, at long last to be discharged. Still, instinctively one does everything possible to postpone the final settlement. Such is the game that Nature plays with us.

> To Gertrud Warschauer, February 5, 1955. Quoted in Nathan and Norden, *Einstein on Peace*, 616. Einstein Archives 39-532

I want to go when I want. It is tasteless to prolong life artificially. I have done my share; it is time to go. I will do it elegantly.

> Quoted by Helen Dukas in her letter to Abraham Pais, an Einstein biographer, April 30, 1955. See Pais, *Subtle Is the Lord*, 477. (Helen Dukas's version in her own account of Einstein's last days is slightly different: "How undignified [to undergo the proposed surgery]. I'm going when I want to go—elegantly!" See Calaprice, *The New Quotable Einstein*, appendix, for an English translation of the account, or Einstein Archives 39-071 for the original German.)

Look into nature, and then you will understand it better.

> Quoted by Margot Einstein in a letter to Carl Seelig, May 8, 1955. Thanks to Barbara Wolff for this new source.

I feel such solidarity with all those alive that it is immaterial to me where the individual begins and where he ends.

> Quoted by Max Born, *Physik im Wandel meiner Zeit* (Wiesbaden, Germany: Vieweg, 1957), 240. First mentioned by Hedi Born in a letter of 1926 or 1927, where she says that Einstein made the remark when he was critically ill.

This house will never become a place of pilgrimage where the pilgrims come to look at the bones of the saint.

> In reply to a student's question on what would become of his house after his death. Recalled by John Wheeler in

French, *Einstein*, 22; also in Wheeler, "Mentor and Sounding Board," in Brockman, ed., *My Einstein*, 35

I want to be cremated so people won't come to worship at my bones.

Quoted by Einstein biographer Abraham Pais, *Manchester Guardian*, December 17, 1994. Most likely this is a paraphrase of Einstein's spoken wishes.

On Education, Students, and
Academic Freedom

An organization cannot in itself engender intellectual activity, but rather can only support what is already in existence.

To Rudolf Lindemann, October 7, 1919, regarding the forming of student associations to replace fraternities and their "barbaric traditions." *CPAE*, Vol. 9, Doc. 125

The inclination of the pupil for a particular profession must not be neglected, especially because such inclination usually asserts itself at an early age, being occasioned by personal gifts, by example of other members of the family, and by various other circumstances.

1920. Quoted by Moszkowski, *Conversations with Einstein*, 65

Most teachers waste their time by asking questions that are intended to discover what a pupil does not know, whereas the true art of questioning is to discover what the pupil does know or is capable of knowing.

Ibid.

*I was astonished at the close friendships between your faculty and students—a condition rarely seen or even possible in German institutions.

From an address at City College of New York, April 21, 1921. Quoted in *The Campus* (CCNY), April 26, 1921, 1, 2. See also Illy, *Albert Meets America*, 114

*Studying, and striving for truth and beauty in general, is a sphere in which we are allowed to be children throughout life.

> Dedication to Adriana Enriques, ca. October 22, 1921. Einstein Archives 36-588

In the matter of physics [education], the first lessons should contain nothing but what is experimental and interesting to see.

> From "Einstein on Education," *Nation and Athenaeum*, December 3, 1921. The unnamed author is quoting from Moszkowski, *Conversations with Einstein*, 69

It is not so very important for a person to learn facts. For that he does not really need college. He can learn them from books. The value of an education in a liberal arts college is not the learning of many facts, but the training of the mind to think something that cannot be learned from textbooks.

> Written in 1921, on Thomas Edison's opinion that a college education is useless. Quoted in Frank, *Einstein: His Life and Times*, 185

It is the supreme art of the teacher to awaken joy in creative expression and knowledge.

> Translation of the quotation on a plaque in the astronomy building of Pasadena City College (the German inscription reads "Es ist die wichtigste Kunst des Lehrers, die Freude am Schaffen und am Erkennen zu wecken"). Einstein dedicated the building and its observatory on February 26,

1931, with a short speech, and also contributed the above words to be inscribed on the small bronze dedicatory plaque inside the building. The elaborate ceremony was attended by most Pasadena schoolchildren as well. See *Pasadena Star News*, February 26, 1931, and *Pasadena Chronicle*, February 27, 1931 (thanks to Dan Haley, Shelley Erwin, and Mane Hakopyan of PCC for these references and copies of the articles). It is not clear if these words are original to Einstein, as Dan Haley, research librarian at Pasadena City College, found a similar statement written by Anatole France in *The Crime of Sylvestre Bonnard*, part 2, chapter 4: "The art of teaching is only the art of awakening the natural curiosity of young minds for the purpose of satisfying it afterwards," which, I discovered, dates back to 1881. No doubt other people have had the same thought through the centuries and the concept is not original to France, either. This sentence is also used as the epigram in *Mein Weltbild* (1934), 25, in the reprinting of Einstein's dedication speech; here it is also not clear if these words are Einstein's own or if he is using an unsourced quotation by someone else, but my feeling is they are his own. See also the next quotation, written three years later.

*The most valuable thing a teacher can impart to children is not knowledge and understanding per se but a *longing* for knowledge and understanding, and an appreciation for intellectual values, whether they be artistic, scientific, or moral.

Similar to the preceding quotation, but written for the National Council of Supervisors of Elementary Science in 1934. Einstein Archives 28-277

*Numerous are the academic chairs, but rare are wise and noble teachers. Numerous and large are the lecture halls, but far from numerous the young people who genuinely thirst for truth and justice. Numerous are the wares that nature produces by the dozen, but her choice products are few.

From "On Academic Freedom," April 28, 1931. See Rowe and Schulmann, *Einstein on Politics*, 464. Einstein Archives 28-151

*Today also there is an urge toward social progress, toward tolerance and freedom of thought, toward a larger political unity. . . . But the students at our universities have ceased as completely as their teachers to embody the hopes and ideals of the people.

Ibid.

Specialization in every sphere of intellectual work is producing an ever-widening gulf between the intellectual worker and the non-specialist, which makes it more difficult for the life of the nation to be fertilized and enriched by the achievements of art and science.

From "Congratulations to Dr. Solf," October 25, 1932. Reprinted in *The World as I See It*, 20

*There are certain occupations, even in modern society, which entail living in isolation and do not require great physical or intellectual effort. Such occupations as the service of lighthouses and lightships come to mind. Would it not be possible to place young people who wish to think about scientific problems, especially of a mathematical or philosophical nature, in such occupations? Very few young people with such ambitions have, even during the most productive period of their lives, the opportunity to devote themselves undisturbed for any length of time to problems of a scientific nature.

From a speech in the Royal Albert Hall, "Science and Civilization," October 3, 1933. Published in 1934 as "Europe's Danger—Europe's Hope." See Rowe and Schulmann, *Einstein on Politics*, 280. Einstein Archives 28-253

Never regard your study as a duty, but as the enviable opportunity to learn the liberating beauty of the intellect for your own personal joy and for the profit of the community to which your later work will belong.

From a statement given to the Princeton University freshman publication, *The Dink*, December 1933. Einstein Archives 28-257

In the teaching of geography and history, a sympathetic understanding [should] be fostered for the characteristics of the different peoples of the world, especially for those whom we are in the habit of describing as "primitive."

From "Education and World Peace," message to the conference of the Progressive Education Association, November 23, 1934. Published in *Progressive Education* 9 (1934), 440; reprinted in *Ideas and Opinions*, 58

In the schools, history should be used as a means of interpreting progress in *civilization*, and not for inculcating ideals of imperialistic power and military success.

Ibid.

Humiliation and mental oppression by ignorant and selfish teachers wreak havoc in the youthful mind that can never be undone and often exert a baleful influence in later life.

From "Paul Ehrenfest in Memoriam." Reprinted in *Out of My Later Years* (1934), 214–217. Einstein Archives 5-136

To me the worst thing seems to be for a school principally to work with the methods of fear, force, and artificial authority. Such treatment destroys the sound sentiments, the sincerity, and the self-confidence of the pupil.

From an address at the celebration of the tercentenary of higher education in America at the State University of New York in Albany, October 15, 1936. In *School and Society* 44 (1936), 589–592. Published as "On Education," in *Ideas and Opinions*, 62. Einstein Archives 29-080

The aim [of education] must be the training of independently acting and thinking individuals who, however, see in the service to the community their highest life achievement.

Ibid., 60

The school should always have as its aim that the young person leave it as a harmonious personality, not as a specialist.

> Ibid., 64. The next quotation, written sixteen years later, reveals what might otherwise happen.

Otherwise, he—with his specialized knowledge—more closely resembles a well-trained dog than a harmoniously developed person.

> From an interview with Benjamin Fine, *New York Times*, October 5, 1952. Reprinted as "Education for Independent Thought" in *Ideas and Opinions*, 66. Einstein Archives 60-723

Freedom of teaching and of opinion in book or press is the foundation for the sound and natural development of any people.

> From an address written for a gathering of university teachers that never took place, 1936. Published as "At a Gathering for Freedom of Opinion" in *Out of My Later Years*, 183–184; Einstein Archives 28-333

The real difficulty, the difficulty that has baffled the sages of all times, is this: how can we make our teaching so potent in the emotional life of man that its influence should withstand the pressure of the elemental psychic forces in the individual?

> From an address at Swarthmore College, June 6, 1938. Published as "Morals and Emotions" in *Out of My Later Years*. Einstein Archives 29-083

*Only understanding for our neighbors, justice in our own dealings, and willingness to help our fellow men can give human society permanence and assure security for the individual. Neither intelligence nor inventions nor institutions can serve as substitutes for these most vital parts of education.

From a CBS radio address for the United Jewish Appeal, March 21, 1939. See also Jerome, *Einstein on Israel and Zionism*, 141. Einstein Archives 28-475

The school of life is chaotic and planless, while the school system operates according to a definite plan. . . . That explains . . . why education is such an important political instrument: there is always the danger that it may become an object of exploitation by contending political groups.

From a message to the New Jersey Education Association, Atlantic City, November 10, 1939. Published in Nathan and Norden, *Einstein on Peace*, 389. Einstein Archives 70-486

*It is a very grave mistake to think that the joy of seeing and searching can be promoted by means of coercion and a sense of duty. . . . I believe that it would be possible to rob even a healthy beast of prey of its voraciouslness . . . with the aid of a whip, to force the beast to devour continuously, even when not hungry.

Written in 1946 for "Autobiographical Notes," 17–19

The crippling of individuals I consider the worst evil of capitalism. Our whole educational system suffers from this evil. An exaggerated competitive attitude is inculcated into the student, who is trained to worship material success as a preparation for his future career.

From "Why Socialism?" *Monthly Review,* May 1949. See Rowe and Schulmann, *Einstein on Politics,* 445

*The education of the individual, in addition to promoting his innate abilities, would attempt to develop in him a sense of responsibility for his fellow men in place of the glorification of power and success in our present society.

On education in a socialist system. Ibid., 445–446

It is in fact nothing short of a miracle that modern methods of instruction have not yet entirely strangled the holy curiosity of inquiry; for this delicate little plant, aside from stimulation, stands mainly in need of freedom; without this it goes to wrack and ruin without fail.

Ibid., 17

Teaching should be such that what is offered is perceived as a valuable gift and not as a hard duty.

From an interview with Benjamin Fine, *New York Times,* October 5, 1952. Reprinted as "Education for Independent

Thought" in *Ideas and Opinions*, 66. Einstein Archives 60-723

I never had the chance to teach youngsters. A pity. I would actually have liked to teach high school.

Quoted by Fantova, "Conversations with Einstein," October 17, 1953

By academic freedom I understand the right to search for truth and to publish and teach what one holds to be true. This right also implies a duty: one must not conceal any part of what one has recognized to be true. It is evident that any restriction of academic freedom acts in such a way as to hamper the dissemination of knowledge among the people and thereby impedes rational judgment and action.

From a statement for a conference of the Emergency Civil Liberties Committee, March 13, 1954. Quoted in Nathan and Norden, *Einstein on Peace*, 551. Facsimile in Cahn, *Einstein*, 97. Einstein Archives 28-1025

I am opposed to examinations—they only deter from the interest in studying. No more than two exams should be given throughout a student's [college] career. I would hold seminars, and if the young people are interested and listen, I would give them a diploma.

Quoted by Fantova, "Conversations with Einstein," January 20, 1955

You should try to remember that a dedicated teacher is a valuable messenger from the past, and can be an escort to your future.

Response to a student who complained about his teacher.
Quoted by Richards, *Einstein as I Knew Him*, Postscript

On and to Friends, Specific Scientists, and Others

On Michele Besso (1873–1955)

Now he has departed from this strange world a little ahead of me. That signifies nothing. For those of us who believe in physics, the distinction between past, present, and future is only a stubbornly persistent illusion.

> On lifelong friend Michele Besso, in a letter of condolence to the Besso family, March 21, 1955, less than a month before his own death. Einstein Archives 7-245

What I admired most in him as a human being is that he managed to live for so many years not only in peace but also in lasting harmony with a woman—an undertaking in which I twice failed rather miserably.

> Ibid.

On Niels Bohr (1885–1962)

Not often in my life has a person given me such joy by his presence as you have. . . . I'm studying your great papers now, and when I get stuck somewhere I have the pleasure of seeing your kindly, boyish face before me, smiling and explaining.

> To Niels Bohr, the Danish physicist and future Nobel laureate (1922), May 2, 1920. *CPAE*, Vol. 10, Doc. 4

Bohr was here, and I'm as enamored of him as you are. He is like an extremely sensitive child who moves around in this world in a sort of trance.

To Paul Ehrenfest, May 4, 1920. *CPAE*, Vol. 10, Doc. 6

What is so marvelously attractive about Bohr as a scientific thinker is his rare blend of boldness and caution; seldom has anyone possessed such an intuitive grasp of hidden things combined with such a strong critical sense. . . . He is unquestionably one of the greatest discoverers of our age in the scientific field.

From "Niels Bohr," February 1922, published in Einstein, *The World as I See It* (1934). Einstein Archives 8-062

He is truly a man of genius. . . . I have full confidence in his way of thinking.

To Paul Ehrenfest, March 23, 1922. Einstein Archives 10-035

*That [Bohr could] discover the major laws of the spectral lines and of the electron-shells of the atoms together with their significance for chemistry appears to me like a miracle. . . . This is the highest form of musicality in the sphere of thought.

Written in 1946 for "Autobiographical Notes," 47

On Max Born (1882–1970)

Born became a pensioner in Edinburgh, and his pension is so small that he couldn't afford to live in England and had to move to Germany.

> On the German physicist whom Einstein admired. Born won the Nobel Prize in 1954, which may have helped his financial circumstances. Quoted by Fantova, "Conversations with Einstein," November 2, 1953

On Louis Brandeis (1856–1941)

I know of no other person who combines such profound intellectual gifts with such self-renunciation while finding the whole meaning of his life in quiet service to the community.

> To Supreme Court Justice Louis Brandeis, November 10, 1936. Einstein Archives 35-046

On Pablo Casals (1876–1973)

What I admire in him particularly is his steadfast demeanor not only against the oppressors of his people, but against all those opportunists who are always ready to make a pact with the devil. He has clearly recognized that the world is more threatened by those who tolerate evil or support it than the evil-doers themselves.

> Written March 30, 1953. Einstein admired the Spanish cel-list not only for his music, but for his humanism and staunch opposition to the fascist Franco regime in Spain. Einstein Archives 34-347

On Charlie Chaplin (1889–1977)

[He] had set up a Japanese theater in his home, with authentic Japanese dances being performed by Japanese girls. Just as in his films, Chaplin is an enchanting person.

> To the Lebach family, January 16, 1931, after visiting the film actor in Hollywood. (Later in the month, Einstein attended the premiere of Chaplin's *City Lights* with him.) Einstein Archives 47-373

*Even Chaplin looks at me like I'm some kind of exotic creature and doesn't know what to make of me. In my room he acted as if he were being brought into a temple.

> Recalled by Konrad Wachsmann, as quoted in Grüning, *Ein Haus für Albert Einstein*, 145

On Marie Curie (1867–1934)

I do not believe Mme Curie is power-hungry or hungry for whatever. She is an unpretentious, honest person with more than her share of responsibilities and burdens. She has a sparkling intelligence, but despite her passionate nature she is not attractive enough to present a danger to anyone.

To Heinrich Zangger, November 7, 1911, regarding the French-Polish physicist and Nobel laureate's (1903, physics; 1911, chemistry) alleged affair with married French physicist Paul Langevin. *CPAE*, Vol. 5, Doc. 303

I am compelled to tell you how much I have come to admire your intellect, your vitality, and your honesty, and that I consider myself fortunate to have made your personal acquaintance in Brussels.

To Marie Curie, November 23, 1911. *CPAE*, Vol. 5, Doc. 312a, embedded in Vol. 8

I am deeply grateful to you and your friends that you so cordially allowed me to participate in your daily life. To witness such marvelous camaraderie among such people is the most uplifting thing I can think of. Everything looked so natural and uncomplicated with you, like a good work of art. . . . I wish to ask your forgiveness if by any chance my crude manners sometimes made you feel uncomfortable.

To Marie Curie, April 3, 1913. *CPAE*, Vol. 5, Doc. 435

Madame Curie is very intelligent but short on emotion [*Häringseele*], meaning that she is deficient in feelings of joy and pain. Almost the only way she expresses her feelings is to rail against things she doesn't like. And she has a daughter who is even worse—like a grenadier. This daughter is also very gifted.

> To Elsa Löwenthal, ca. August 11, 1913. *CPAE*, Vol. 5, Doc. 465

*I would not dare grumble to you in this fashion if I did not think of you as a sister in defiance, one who, somewhere in her soul, has always had an understanding of such feelings, and one to whom I have always felt particularly close.

> To Marie Curie, December 25, 1923, while complaining to her about the League of Nations and giving his reasons for resigning from its Committee on Intellectual Cooperation. He rejoined six months later. See also Rowe and Schulmann, *Einstein on Politics*, 196. Einstein Archives 8-431

Her strength, her purity of will, her austerity toward herself, her objectivity, her incorruptible judgment— all these were of a kind seldom found in a single individual. . . . Once she had recognized a certain way as a right one, she pursued it without compromise and with extreme tenacity.

> At the Curie memorial celebration, Roerich Museum, New York, November 23, 1934. Einstein Archives 5-142

On Paul Ehrenfest (1880–1933)

His sense of inadequacy, objectively unjustified, plagued him incessantly, often robbing him of the peace of mind necessary for tranquil research. . . . His tragedy lay precisely in an almost morbid lack of self-confidence. . . . The strongest relationship in his life was that toward his wife and fellow worker . . . his intellectual equal. . . . He repaid her with a veneration and love such as I have not often witnessed in my life.

From "Paul Ehrenfest in Memoriam," written after the physicist and close friend's suicide, 1934. Ehrenfest had shot his sixteen-year-old son, Vassik, who had Down syndrome, in the waiting room of the institute where he was being treated. Then he shot himself. Reprinted in *Out of My Later Years*, 214–217. Einstein Archives 5-136

He was not only the best teacher in our profession whom I have ever known; he was also passionately preoccupied with the development and destiny of people, especially his students. To understand others, to gain their friendship and trust, to aid anyone embroiled in outer or inner struggles, to encourage youthful talent—all this was his real element, almost more than his immersion in scientific problems.

Ibid.

On Dwight D. Eisenhower (1890–1969)

Eisenhower said on the radio, "One can't win peace with military might." I found this remarkable.

Eisenhower is good—he is advising Americans not to get mixed up in Chinese politics.

> On the thirty-fourth U.S. president (served 1953–1961).
> Quoted by Fantova, "Conversations with Einstein," November 13, 1953

On Michael Faraday (1791–1867)

This man loved mysterious Nature as a lover loves his distant beloved. . . . [In Faraday's day] there did not yet exist the dull specialization that, through horn-rimmed glasses and arrogance, destroys the poetry.

> On the English chemist and physicist. To Gertrud Warschauer, December 27, 1952. Einstein Archives 39-517

On Abraham Flexner (1866–1959)

Flexner, the former director of the Institute, is one of the few enemies I have here. Years ago I conducted a revolt against him, from which he fled.

> Quoted by Fantova, "Conversations with Einstein," January 23, 1954. Flexner was director of the Institute for Advanced Study when Einstein was hired. He was so protective and possessive of Einstein that he purposely did not tell him that President Roosevelt had invited him to the White House through the offices of the Institute. Einstein eventually got wind of it, apologized to Roosevelt, and received another invitation, which he accepted.

On or to Sigmund Freud (1856–1939)

Why do you emphasize happiness in my case? You, who have gotten under the skin of so many people and, indeed, of humanity, have had no occasion to slip under mine.

To Sigmund Freud, March 22, 1929, in reply to the Viennese psychoanalyst's letter on Einstein's fiftieth birthday, in which he congratulated him for being a "happy one." Einstein Archives 32-530

I am not prepared to accept all his conclusions, but I consider his work an immensely valuable contribution to the science of human behavior. I think he is even greater as a writer than as a psychologist. Freud's brilliant style is unsurpassed by anyone since Schopenhauer.

From an interview with G. S. Viereck, "What Life Means to Einstein," *Saturday Evening Post,* October 26, 1929; reprinted in Viereck, *Glimpses of the Great,* 443

*I understand Jung's vague, imprecise notions, but I consider them worthless; a lot of talk without any clear direction. If there has to be a psychiatrist, I should prefer Freud. I do not believe in him, but I love very much his concise style and his original, although rather extravagant, mind.

Diary entry of December 6, 1931. See Nathan and Norden, *Einstein on Peace,* 185. Later, after a period of doubt, he even expressed belief in Freud's ideas, calling it a blessing in a letter to Freud on April 21, 1936: "It is always a blessing when a great and beautiful conception is proven to be in harmony with reality." See Rowe and Schulmann, *Einstein on Politics,* 220

*You have earned my gratitude and the gratitude of all men for having devoted all your strength to the search for truth and for having shown the rarest courage in professing your convictions all your life.

To Freud, December 3, 1932. Einstein Archives 32-554

The old one ... had a sharp vision; no illusions lulled him to sleep except for an often exaggerated faith in his own ideas.

To A. Bacharach, July 25, 1949. Einstein Archives 57-629

On Galileo Galilei (1564–1642)

*His aim was to substitute for a petrified and barren system of ideas the unbiased and strenuous quest for a deeper and more consistent comprehension of physical and astronomical facts.

From the foreword to *Galileo Galilei: Dialogue concerning the Two Chief World Systems,* ca. 1953. See also Rowe and Schulmann, *Einstein on Politics,* 133. Einstein Archives 1-174

That, alas, is vanity. You find it in so many scientists. You know, it has always hurt me to think that Galilei did not acknowledge the work of Kepler.

From I. Bernard Cohen, "Einstein's Last Interview," April 1955, shortly before Einstein's death. Published in *Scientific American* 193, no. 1 (July 1955), 69; reprinted in Robinson, *Einstein*, 215

The discovery and use of scientific reasoning by Galileo was one of the most important achievements in the history of human thought, and marks the real beginning of physics.

Einstein and Infeld, *Evolution of Physics*, 7

On Mahatma Gandhi (1869–1948)

I admire Gandhi greatly but I believe there are two weaknesses in his program. Nonresistance is the most intelligent way to face difficulty, but it can be practiced only under ideal conditions. . . . It could not be carried out against the Nazi Party today. Then, Gandhi makes a mistake in trying to abolish the machine from modern civilization. It is here, and it must be dealt with.

From an interview, *Survey Graphic* 24 (August 1935), 384, 413, discussing the Indian pacifist's goals

A leader of his people, unsupported by any outward authority: a politician whose success rests not upon craft or the mastery of technical devices, but simply on the convincing power of his personality; a victorious fighter who has always scorned the use of force; a man of wisdom and humility, armed with resolve and inflexible consistency, who has devoted all his strength to the uplifting of his people and the betterment of their lot; a man who has confronted the brutality of Europe with the dignity of the simple human being, and thus at all times risen superior. Generations to come, it may well be, will scarce believe that such a one as this ever in flesh and blood walked upon this Earth.

Statement on the occasion of Gandhi's seventy-fifth birthday, 1939. In Gandhiji, *Gandhiji: His Life and Work* (Bombay: Karnatak, 1944), xi. Reprinted in *Einstein on Humanism*, 94. Einstein Archives 28-60

I believe that Gandhi's views were the most enlightened among all of the political men of our time. We should strive to do things in his spirit; not to use violence in fighting for our cause, but by nonparticipation in what we believe is evil.

From a United Nations Radio interview, June 16, 1950, recorded in the study of Einstein's home in Princeton. Reprinted in the *New York Times*, June 19, 1950; also quoted in Pais, *Einstein Lived Here*, 110

Gandhi, the greatest political genius of our time, indicated the path to be taken. He gave proof of what sacrifice man is capable once he has discovered the right path. His work on behalf of India's liberation is living testimony to the fact that man's will, sustained by an indomitable conviction, is more powerful than material forces that seem insurmountable.

> To the editor of the Japanese magazine *Kaizo*, September 20, 1952. See Rowe and Schulmann, *Einstein on Politics*, 489. Einstein Archives 60-039

Gandhi's development resulted from extraordinary intellectual and moral forces in combination with political ingenuity and a unique situation. I think Gandhi would have been Gandhi even without Thoreau and Tolstoy.

> To Walter Harding, a member of the Thoreau Society, August 19, 1953. Quoted in Nathan and Norden, *Einstein on Peace*, 594. Einstein Archives 32-616

On Johann Wolfgang von Goethe (1749–1832)

I feel in him a certain condescending attitude toward the reader, and miss the humility that is comforting, especially when it comes from great men.

> On the German poet. To Leopold Casper, April 9, 1932. Einstein Archives 49-380

I admire Goethe as a poet without peer, and as one of the smartest and wisest men of all time. Even his scholarly ideas deserve to be held in high esteem, and his faults are those of any great man.

> Ibid.

On Fritz Haber (1868–1934)

*Haber's picture can be seen everywhere, unfortunately. It hurts me every time I think about it. I must reconcile myself to the idea that this man who is otherwise so outstanding has succumbed to personal vanity . . . of the most untasteful kind. Such inflated vanity is typical of Berliners. What a difference between them and the French and English!

> To Elsa Einstein, December 2, 1913. *CPAE*, Vol. 5, Doc. 489. Haber's fervent German patriotism, around the time he converted from Judaism to Christianity, was distasteful to Einstein. Haber, a chemist who would win the Nobel Prize for 1918, worked with poison gases to be used in warfare and personally oversaw the first use of chlorine in trench warfare in April 1915. His wife, also a chemist, disapproved and committed suicide three weeks later.

On Lord Richard B. S. Haldane (1856–1928)

*This was the first time I've ever heard of such an important man who speaks at least briefly with his mother every day. My scientific discourse with Lord Haldane is for me a source of great excitement, and knowing him personally is a meaningful experience.

> To Haldane's mother, Mary, after a trip to Britain, June 15, 1921. Haldane was a prominent British Labour politician, lawyer, and philosopher. *CPAE*, Vol. 12, Doc. 149

On Werner Heisenberg (1901–1976)

Professor Heisenberg was here, a German. He was an important Nazi (*ein grosser Nazi*). He is a great physicist, but not a very pleasant man.

On the Nobel laureate (1932) and creator of quantum mechanics. Quoted by Fantova, "Conversations with Einstein," October 30, 1954

On Adolf Hitler (1889–1945)

I am very happy here and enjoying the American summer as well as the news about Hitler's mad deed of desperation. He and his henchmen will be unable to carry on for a long time after he has destroyed his powerful tool and his halo. Then a general will take over and the Jews will have some breathing room.

From an optimistic letter to Rabbi Stephen Wise, July 3, 1934. "Hitler's mad deed of desperation" was the Night of Long Knives in late June, during which he arrested top Stormtrooper (SA) leaders whom he suspected of being disloyal to him, though there was no such evidence. The SA leader, Ernst Röhm, was shot and others were bludgeoned to death. Röhm had been Hitler's "tool" in fighting Communists in the late 1920s. The general was probably Kurt von Schleicher, Hitler's immediate predecessor, though at this point Einstein probably did not know that he had been executed as well. Einstein Archives 35-152

Hitler appeared, a man with limited intellectual abilities and unfit for any useful work, bursting with envy and bitterness against all whom circumstance and nature had favored over him. . . . He picked up this human flotsam on the streets and in the taverns and organized them around himself. This is the way he launched his political career.

> From an unpublished manuscript, 1935; published later in compilations: Nathan and Norden, *Einstein on Peace*, 263–264; Dukas and Hoffman, *Albert Einstein, the Human Side*, 110; Rowe and Schulmann, *Einstein on Politics*, 295. Einstein Archives 28-322

*I haven't forgotten that the Swiss authorities didn't stand by me in any way when Hitler stole all of my savings, even those designated for my children.

> To Heinrich Zangger, September 18, 1938. Einstein Archives 40-116

Yes, my girlfriends and sailboat remained in Berlin. But Hitler only wanted the latter, which was insulting to the former.

> Recalled by his assistant Ernst Straus, in Seelig, *Helle Zeit, Dunkle Zeit*, 68

On Heike Kamerlingh Onnes (1853–1926)

A life has ended that will always remain a role model for future generations. . . . No other person have I known for whom duty and joy were one and the same. This was the reason for his harmonious life.

To the widow of the Dutch physicist and Nobel laureate (1913), February 25, 1926. Einstein Archives 14-389

On Immanuel Kant (1724–1804)

*Kant's much-praised view on Time reminds me of Andersen's tale of the emperor's new clothes, only that instead of the emperor's new clothes we have the form of intuition.

To his student Ilse [Rosenthal-]Schneider, September 15, 1919, on the eighteenth-century Prussian philosopher. Einstein Archives 20-261

*Kant is sort of a highway with lots and lots of milestones. Then all the little dogs come and each deposits his contribution at the milestones.

Said to Ilse [Rosenthal-]Schneider, around the same time as above. See Rosenthal-Schneider, *Reality and Scientific Truth*, 90

What seems to me the most important thing in Kant's philosophy is that it speaks of a priori concepts for the construction of science.

At a discussion in the Société Française de Philosophie, July 1922. Quoted in *Bulletin Société Française de Philosophie* 22 (1922), 91; reprinted in *Nature* 112 (1923), 253

If Kant had known what is known to us today of the natural order, I am certain that he would have fundamentally revised his philosophical conclusions. Kant built his structure upon the foundations of the world outlook of Kepler and Newton. Now that the foundation has been undermined, the structure no longer stands.

From an interview with Chaim Tchernowitz, *The Sentinel*, date unknown

Kant, thoroughly convinced of the indispensability of certain concepts, took them—just as they are selected—to be the necessary premises for every kind of thinking and differentiated them from concepts of empirical origin.

Written in 1946 for "Autobiographical Notes," 13

On George Kennan (1904–2005)

Princeton University Press sent me George Kennan's new book [*Realities of American Foreign Policy*] and I read it right away. I liked it very much. Kennan has done his job well.

Quoted by Fantova, "Conversations with Einstein," August 22, 1954. Kennan, a Princeton resident, was ambassador to the Soviet Union and developer of the foreign policy of containment.

On Johannes Kepler (1571–1630)

[Kepler] belonged to those few who cannot do otherwise than openly acknowledge their convictions on every subject. . . . [His] lifework was possible only when he succeeded in freeing himself to a large extent from the spiritual tradition in which he was born. . . . He does not speak about this, but the inner struggle is reflected in his letters.

> On the Swabian astronomer and mathematician. From Introduction to Carola Baumgardt, *Johannes Kepler: Life and Letters* (New York: Philosophical Library, 1951), 12–13

Neither by poverty, nor by incomprehension of the contemporaries who ruled over his life and work, did he allow himself to be crippled or discouraged.

> Ibid.

There we meet a finely sensitive person, passionately dedicated to the search for a deeper insight into the essence of natural events, who, despite internal and external difficulties, reached his loftily placed goal.

> Ibid., 9

To Lover and Alleged Soviet Spy, Margarita Konenkova (c. 1900–?)

In 1998, a group of letters, written by Einstein to a mysterious woman with whom he had had an

affair before and during World War II while living
in Princeton, was auctioned at Sotheby's in New
York. The enigmatic woman was Margarita Konenkova. According to a book published in 1995 by
former Soviet spymaster Pavel Sudoplatov, Konenkova was a Russian agent whose official mission
was to introduce Einstein to the Soviet vice-consul
in New York and "to influence Oppenheimer and
other prominent American scientists whom she frequently met in Princeton." She did succeed in introducing Einstein to the Soviet vice-consul, Pavel
Mikhailev, and Einstein refers to him in their letters.
But Sudoplatov's account otherwise becomes
suspect because, among other things, he places
Oppenheimer in Princeton at the time, although
he was actually 2,000 miles away in Los Alamos,
New Mexico, helping to design the bomb; he did
not come to Princeton until 1947, two years after
Mrs. Konenkova had left. Einstein was sixty-six
at the time the letters started in late 1945, and
Mrs. Konenkova, whose husband Sergei created
the bronze bust of Einstein at the Institute for Advanced Study in 1935, was in her mid-forties
(though the *New York Times* put her age at 51). She
and Sergei were Russian emigrés who lived in
Greenwich Village from the early 1920s to 1945,
when they were recalled to the Soviet Union. Margarita was a good friend of Margot Einstein, whose
ex-husband had been an attaché at the Soviet
Embassy in Berlin in the early 1930s. She was
known to have had affairs with other prominent
men, and there is no indication that Einstein was
aware that she might be a spy. See the *New York
Times*, June 1, 1998, A1, and Sotheby's Catalog,
June 26, 1998; all the letters below are in Sotheby's

Catalog or in the *New York Times*, slightly differently translated.

Because of your great love for your homeland, you would probably have become bitter in time if you had not taken this step [to return to Russia]. For unlike me, you have decades of active work and life ahead of you, whereas for me everything indicates . . . that my days will have run their course before too long. I think of you often.

Letter of November 8, 1945

I recently washed my hair by myself, but not with great success; I am not as careful as you are. But everything here reminds me of you: . . . the dictionaries, the wonderful pipe which we thought was lost, and all the other little things in my hermit's cell; and also my lonely nest.

Letter of November 27, 1945. It appears that women liked to play with Einstein's famous hair. Another woman friend in Princeton is known to have cut his hair (obviously not frequently enough).

People are living now [after the war] just as they were before . . . and it is clear that they have learned nothing from the horrors they have had to deal with. The little intrigues with which they had complicated their lives before are again taking up most of their thoughts. What a strange species we are.

Letter of December 30, 1945

With best wishes and kisses, if this letter reaches you. And may the devil take anyone who intercepts it.

Letter of February 8, 1946

I can imagine that the May Day ceremonies [in Moscow] must have been marvelous. But you know that I watch these exaggerated patriotic exhibitions with concern. I always try to convince people of the importance of cosmopolitan, reasonable, and fair thinking.

Letter of June 1, 1946

On Paul Langevin (1872–1946)

If he loves Mme Curie and she loves him, they do not have to run off together, because they have plenty of opportunities to meet in Paris. But I don't have the impression that anything special is going on between the two of them; rather, I found all three of them bound by a pleasant and innocent relationship.

To Heinrich Zangger, November 7, 1911, on the married French physicist's rumored affair with Marie Curie. The third person Einstein is referring to was Jean Perrin, another colleague. *CPAE*, Vol. 5, Doc. 303

There are so very few in any one generation in whom clear insight into the nature of things is joined with an intense feeling for the challenge of true humanity and the capacity for militant action. When such a man departs, he leaves a gap that seems unbearable to his survivors. . . . His desire to promote a happier life for all men was perhaps even stronger than his craving for pure intellectual enlightenment. No one who appealed to his social conscience ever went away empty-handed.

From the obituary for Langevin in *La Pensée*, n.s., no. 12 (May–June 1947), 13–14. Einstein Archives 5-150

I had already heard of Langevin's death. He was one of my dearest acquaintances, a true saint, and talented besides. It is true that the politicians exploited his goodness, for he was unable to see through the base motives that were so foreign to his nature.

To Maurice Solovine, April 9, 1947. Published in *Letters to Solovine*, 99. Einstein Archives 21-250

On Max von Laue (1879–1960)

*It was interesting to see the way in which [von Laue] cut himself off, step by step, from the traditions of the herd, under the influence of a strong sense of justice.

On his friend and German Nobel laureate in physics. Written to Max Born, September 7, 1944, in *Born-Einstein Letters*, 145. Einstein Archives 8-207

On Philipp Lenard (1862–1947)

I admire Lenard as a master of experimental physics; however, he has yet to accomplish anything of importance in theoretical physics, and his objections to the general theory of relativity are so superficial that I had not deemed it necessary until now to reply to them in detail.

On the staunch Nazi, anti-Semite, and Nobel laureate in physics (1905). *Berliner Tageblatt*, August 27, 1920, 1–2. *CPAE*, Vol. 7, Doc. 45

On Vladimir Ilyich Lenin (1870–1924) and Friedrich Engels (1820–1895)

I respect Lenin as a man who gave all his energy, at a total sacrifice of his personal life, to dedicating himself to the realization of socialist justice. I don't consider his methods appropriate. But one thing is certain: men such as he are the guardians and renewers of mankind's conscience.

Statement for the League of Human Rights on Lenin's death, January 6, 1929. Einstein Archives 34-439

Outside Russia, Lenin and Engels are of course not valued as scientific thinkers and no one would be interested in refuting them as such. The same might also be the case in Russia, except there one doesn't dare say so.

> To K. R. Leistner, September 8, 1932. Einstein Archives 50-877

On Hendrik Antoon Lorentz (1853–1928)

Lorentz is a marvel of intelligence and exquisite tact. A living work of art! In my opinion he was the most intelligent of the theorists present [at the Solvay Congress in Brussels].

> To Heinrich Zangger, November 1911, on the Dutch physicist and Nobel laureate (1902). *CPAE*, Vol. 5, Doc. 305

My feeling of intellectual inferiority with regard to you cannot spoil the great delight of [our] conversation, especially because the fatherly kindness you show to all people does not allow any feeling of despondency to arise.

> To Lorentz, February 18, 1912. *CPAE*, Vol. 5, Doc. 360

*How often when the human affairs surrounding me seemed hopelessly sad, have I found deep consolation in your noble and outstanding personality. A man like you gives consolation and exaltation by his mere existence and example.

> To Lorentz, July 15, 1923. Einstein Archives 16-552

He shaped his life like a precious work of art down to the smallest detail. His never-failing kindness and generosity and his sense of justice, coupled with a sure and intuitive understanding of people and human affairs, made him a leader in any sphere he entered.

From the eulogy at the grave of Lorentz, 1928. Published in *Ideas and Opinions*, 73. Einstein Archives 16-126

To me personally, he meant more than all the others encountered in my lifetime.

From a message for commemoration in Leyden, February 27, 1953. In the enlarged edition of *Mein Weltbild*, 31. Einstein Archives 16-631

On Ernst Mach (1838–1916)

In him, the immediate pleasure gained in seeing and comprehending—Spinoza's *amor dei intellectualis*—was so strong that he looked at the world with the curious eyes of a child until well into old age, so that he could find joy and contentment in understanding how everything is connected.

From the eulogy for the philosopher whose critique of Newton played a role in Einstein's development of relativity theory, even though Mach himself was critical of the theory. In *Physikalische Zeitschrift*, April 1, 1916; *CPAE*, Vol. 6, Doc. 29

Mach was as good a scholar of mechanics as he was a deplorable philosopher.

Quoted in Bulletin *Société Française de Philosophie* 22 (1922), 91; reprinted in *Nature* 112 (1923), 253; see also *CPAE*, Vol. 6, Doc. 29, n. 6, and Vol. 13, "Discussion Remarks"

On Lise Meitner (1878–1968)

*She knows her way around the family of radioactive substances better than I know the way around my own family.

Quoted in Rosenthal-Schneider, *Reality and Scientific Truth*, 113

On Albert Michelson (1852–1931)

I always think of Michelson as the artist in science. His greatest joy seemed to come from the beauty of the experiment itself, and the elegance of the method employed.

From "In Memory of Albert A. Michelson on His 100th Birthday," December 19, 1952, on the physicist and Nobel laureate (1907) who, with Edward Morley in 1881, had already experimentally validated Einstein's postulation that the speed of light is independent of the frame of reference in which it is measured. Einstein Archives 1-168. Einstein said that he was unaware of the experiment when he wrote his 1905 paper on the special theory of relativity; see discussion in Fölsing, *Albert Einstein*, 217–219.

*My admiration for Michelson's experiment is for the ingenious method to compare the location of the interference pattern with the location of the image of the light source. In this way he overcame the difficulty that we are not able to change the direction of the earth's rotation.

To Robert Shankland, September 17, 1953. Einstein Archives 17-203

On Jawaharlal Nehru (1889–1964)

May I tell you of the deep emotion with which I read recently that the Indian Constituent Assembly has abolished untouchability? I know how large a part you have played in the various phases of India's struggle for emancipation, and how grateful lovers of freedom must be to you, as well as to your great teacher Mahatma Gandhi.

To the Indian prime minister (1947–1964). Letter of June 13, 1947. Einstein Archives 32-725

On Isaac Newton (1643–1727)

His clear and wide-ranging ideas will retain their unique significance for all time as the foundation of our whole modern conceptual structure in the sphere of natural philosophy.

From "What Is the Theory of Relativity?" *The Times* (London), November 28, 1919. *CPAE*, Vol. 7, Doc. 25

It was highly honorable of his logical conscience that Newton decided to create absolute space. . . . He could just as well have called the absolute space the "rigid ether." He needed such a reality in order to give objective meaning to acceleration. Later attempts to do without this absolute space in mechanics were (with the exception of Mach's) only a game of hide-and-seek.

To Moritz Schlick, June 30, 1920. *CPAE*, Vol. 10, Doc. 67

In my opinion, the greatest creative geniuses are Galileo and Newton, whom I regard in a certain sense as forming a unity. And in this unity Newton is [the one] who has achieved the most imposing feat in the realm of science.

1920. Quoted by Moszkowski, *Conversations with Einstein*, 40

*He is a yet more significant figure than his own mastery makes him, since he was placed by fate at a turning point in the world's intellectual development. This is brought home vividly to us when we recall that before Newton there was no comprehensive system of physical causality which could in any way render the deeper characters of the world of concrete experience.

From "Isaac Newton," in *Manchester Guardian Weekly*, March 19, 1927, celebrating the second centenary of Newton's death. Reprinted in *Smithsonian Annual Report for 1927*

In one person he combined the experimenter, the theorist, the mechanic, and, not the least, the artist of exposition.

From the Foreword to Newton, *Opticks* (New York: McGraw-Hill, 1931). Einstein Archives 4-046

Newton was the first to succeed in finding a clearly formulated basis from which he could deduce a wide field of phenomena by means of mathematical thinking—logically, quantitatively, and in harmony with experience.

On Newton's upcoming 300th birthday, in *Manchester Guardian*, Christmas 1942. See also *Out of My Later Years*, 201

*To think of him is to think of his work. For such a man can be understood only by thinking of him as a scene on which the struggle for eternal truth took place.

Ibid.

Newton . . . you found the only way which, in your age, was just about possible for a man of highest thought and creative power. The concepts, which you created, are even today still guiding our thinking in physics, although we now know that they will have to be replaced . . . if we aim at a profounder understanding of relationships.

Written in 1946 for "Autobiographical Notes," 31–33

On Alfred Nobel (1833–1896)

Alfred Nobel invented the most powerful explosive ever known up to his time, a means of destruction par excellence. In order to atone for this, in order to relieve his conscience, he instituted his awards for the promotion of peace and for the achievement of peace.

> From "The War Is Won but the Peace Is Not," speech at the fifth Nobel anniversary dinner in New York, December 10, 1945. See Rowe and Schulmann, *Einstein on Politics*, 381–382

On Emmy Noether (1882–1935)

On receiving the new work from Fräulein Noether, I again find it a great injustice that she cannot lecture officially. I would be very much in favor of taking energetic steps in the process [to overturn this rule].

> To Felix Klein, December 27, 1918, on the brilliant German mathematician, Emmy Noether, who was not allowed to be on the faculty of the University of Göttingen because of her gender. *CPAE*, Vol. 8, Doc. 677

It would not have done the Old Guard at Göttingen any harm, had they learned a thing or two from her. She certainly knows what she is doing.

> To David Hilbert, May 24, 1918. *CPAE*, Vol. 8, Doc. 548

In the judgment of the most competent living mathematicians, Fräulein Noether was the most significant creative mathematical genius thus far produced since the higher education of women began.

From the obituary for Emmy Noether, *New York Times*, May 4, 1935. Einstein Archives 5-138

On J. Robert Oppenheimer (1904–1967)

I have to say that Oppenheimer is an exceptional person. Seldom has someone been so talented and upstanding. He may not have contributed anything extraordinary to science, that is, he didn't move science forward, but he is technically very gifted. In all my dealings with him, he has acted with decorum.

On the American physicist, director of the Manhattan Project's Los Alamos Laboratory (1942–1945), and director of the Institute for Advanced Study (1947–1966). Quoted by Fantova, "Conversations with Einstein," April 24, 1954

On Wolfgang Pauli (1900–1958)

This Pauli is a well-oiled head.

On the Austrian physicist and Nobel laureate (1945). To his sister, Maja, August 1933. Einstein Archives 29-416. In a letter to Max and Hedi Born of December 30, 1921, he calls Pauli a "splendid fellow for his 21 years" after Pauli wrote a well-received encyclopedia article summarizing relativity theory. See Born, *Born-Einstein Letters*, 62. *CPAE*, Vol. 12, Doc. 345

On Max Planck (1858–1947)

It was largely because of the decisive and cordial manner in which he supported this theory that it attracted notice so quickly among my colleagues in the field.

> On the German physicist and Nobel laureate (1918) whom Einstein admired, speaking about the special theory of relativity, in "Max Planck as Scientist" (1913). Planck was more responsible than anyone else for firmly establishing relativity theory after 1905. *CPAE*, Vol. 4, Doc. 23

Luring Planck away from here is totally inconceivable. He is rooted to his native land with every fiber of his body, like no other.

> To Heinrich Zangger, June 1, 1919. *CPAE*, Vol. 9, Doc. 52

Planck's misfortune stirs my heart deeply. I could not hold back my tears when I visited him after my return from Rostock. He is wonderfully brave and staid, but his gnawing pain shows through.

> To Max Born, December 8, 1919. *CPAE*, Vol. 9, Doc. 198. Planck's wife had died at age forty-eight in 1909, and his twin daughters died in 1917 and 1919 (both in childbirth). One of his sons, Erwin, had been taken prisoner of war in 1914 by the French and was later released, and the other died in action in 1917. Erwin was hanged by the Nazis in

1945 after his participation in a failed attempt to assassi-
nate Hitler.

How different, and how much better, it would be for
mankind if there were more like him among us. But
it seems not possible. Honorable persons in every
age and everywhere have remained isolated, unable
to influence external events.

To the second Frau Planck, November 10, 1947, about her
husband. Einstein Archives 19-406

He was one of the finest people I have ever
known . . . but he really did not understand physics,
[because] during the eclipse of 1919 he stayed up all
night to see if it would confirm the bending of light
by the gravitational field. If he had really understood the general theory of relativity, he would have
gone to bed the way I did.

Quoted by Ernst Straus in French, *Einstein: A Centenary
Volume*, 31

On Walther Rathenau (1867–1922)

It is easy to be an idealist when one lives in the
clouds. But he was an idealist who lived down on
Earth and knew its scents like few others.

On the German foreign minister who was assassinated in
July 1922 by members of a secret fascist terrorist organization called "Organisation Consul." From obituary for

Rathenau in *Neue Rundschau* 33, no. 8 (1922), 815–816. Einstein Archives 32-819

On several occasions I spent hours in Rathenau's company discussing diverse subjects. These talks tended to be rather one-sided: on the whole, he spoke and I listened. For one thing, it was not so easy to get the floor; for another, it was so pleasant to listen to him that one did not try very hard.

To Johanon Twersky, February 2, 1943. Quoted in Nathan and Norden, *Einstein on Peace*, 52. Einstein Archives 32-836

On Romain Rolland (1866–1944)

He is right in attacking individual greed and the national scramble for wealth that make war inevitable. He may not be far wrong in turning to social revolution as the only means of breaking the war system.

From an interview, *Survey Graphic* 24 (August 1935), 384, 413, on the most prominent pacifist of the time and Nobel laureate in literature (1915)

On Franklin D. Roosevelt (1882–1945)

*The burden he carried was heavy, but his sense of humor gave him an inner freedom seldom found among those who are constantly faced with the most critical decisions. He was unbelievably bound and determined to attain his final goals, yet amazingly

flexible in overcoming the strong resistance any farsighted statesman faces in a democratic country.

> Commemorative words for FDR, *Aufbau* 11, no. 17 (April 27,
> 1945), 7

No matter when this man might have been taken
from us, we would have felt we had suffered an ir-
replaceable loss. . . . May he have a lasting influence
on our thoughts and convictions.

> Ibid. According to the *New York Times*, August 19, 1946,
> Einstein was sure that FDR would have forbidden the
> bombing of Hiroshima had he been alive. Einstein had
> written a letter to FDR in March 1945 warning him of the
> bomb's devastating effects; the president died before he
> had a chance to read it.

I'm so sorry that Roosevelt is president—otherwise
I would visit him more often.

> To friend Frieda Bucky. Quoted in *The Jewish Quarterly* 15,
> no. 4 (Winter 1967–68), 34

On Ilse Rosenthal-Schneider (1891–1990)

*She has a well-founded knowledge in these mat-
ters and is quite independent and original in her
views and opinions. I am convinced she is able to
present philosophy and history of science in an attractive and stimulating manner.

On writing a letter of recommendation for his former Berlin student to fill a permanent lectureship in philosophy of science at the University of Sydney, 1944. See Rosenthal-Schneider, *Reality and Scientific Truth*, 22

On Bertrand Russell (1872–1970)

The clarity, certainty, and impartiality you apply to the logical, philosophical, and human issues in your books are unparalleled in our generation.

To Bertrand Russell, October 14, 1931. See Grüning, *Ein Haus für Albert Einstein*, 369. Einstein Archives 33-155

Great spirits have always encountered opposition from mediocre minds. The mediocre mind is incapable of understanding the man who refuses to bow blindly to conventional prejudices and chooses instead to express his opinions courageously and honestly.

On the controversy surrounding Russell's appointment to the faculty of the City University of New York. Some conservative religious and so-called patriotic New Yorkers regarded Russell as a propagandist against religion and morality and brought legal suit against his appointment. His teaching contract was rescinded. Quoted in the *New York Times*, March 19, 1940. Einstein Archives 33-168

I read [to some visitors] Bertrand Russell's article on religion. I consider him the best of the living writers.

The article is masterfully crafted—everything makes sense.

> Quoted by Fantova, "Conversations with Einstein," December 31, 1953. The article was "What Is an Agnostic," in Leo Rosten's *Religions in America* (1952)

On Albert Schweitzer (1875–1965)

He is a great figure who bids for the moral leadership of the world.

> From an interview, about the Alsatian-born doctor, missionary, humanitarian, and winner of the Nobel Peace Prize (1952). *Survey Graphic* 24 (August 1935), 384, 413

He is the only Westerner who has had a moral effect on this generation comparable to Gandhi's. As in the case of Gandhi, the extent of this effect is overwhelmingly due to the example he gave by his own life's work.

> From a statement originally intended for a new edition of *Mein Weltbild* in 1953 but not published there. Quoted in Sayen, *Einstein in America*, 296. Einstein Archives 33-223

On George Bernard Shaw (1856–1950)

He is a magnificent fellow, with deep insight into the human condition.

On the British writer and winner of the Nobel Prize in literature (1925). To Michele Besso, January 5, 1929, in a discussion of Shaw's book on socialism. See Einstein, *Correspondance avec Michèle Besso 1903–1955* (Paris: Hermann, 1979), 240

Shaw is undoubtedly one of the world's greatest figures. I once said of him that his plays remind me of Mozart. There is not one superfluous word in Shaw's prose, just as there is not one superfluous note in Mozart's music.

In *Cosmic Religion* (1931), 109. Original source unknown.

Thus speaks the Voltaire of our day.

Recalled by Hedwig Fischer, 1928, in P. de Mendelsohn's *The S. Fischer Verlag* (1970), 1164

On Bishop Fulton J. Sheen (1895–1979)

Bishop Sheen is one of the most intelligent [*gescheitesten*] people in today's world. He wrote a book in which he defends religion against science.

On the American Catholic bishop. Quoted by Fantova, "Conversations with Einstein," December 13, 1953. The book Einstein refers to is *Religion without God* (1928). Sheen, on the other hand, disparaged Einstein's "cosmic religion." See "Others on Einstein."

On Upton Sinclair (1878–1968)

He is in the doghouse here because he relentlessly sheds light on the hurly-burly dark side of American life.

On the American writer and Pulitzer Prize winner (1942). To the Lebach family, January 16, 1931. Einstein Archives 47-373

On Baruch Spinoza (1632–1677)

I am fascinated by Spinoza's pantheism. I admire even more his contribution to modern thought . . . because he is the first philosopher who deals with the soul and the body as one, not as two separate things.

From an interview with G. S. Viereck, "What Life Means to Einstein," *Saturday Evening Post*, October 26, 1929, on the seventeenth-century Jewish philosopher, whose thought greatly influenced Einstein; reprinted in Viereck, *Glimpses of the Great*, 448. The German Romantic poet Novalis called Spinoza a "God-intoxicated man."

In my opinion, his opinions have not gained general acceptance by all those striving for clarity and logical rigor only because they require not only consistency of thought but also unusual integrity, magnanimity—and modesty.

To D. Runes, September 8, 1932. Einstein Archives 33-286

*I have been reading Spinoza's letters. He well understood the liberation one gains through pastoral seclusion.

To Leo Szilard, September 15, 1928. Einstein Archives 33-271

Spinoza is one of the deepest and purest souls our Jewish people has produced.

In a letter of 1946. As quoted in Hoffmann, *Albert Einstein: Creator and Rebel*, 95, though a search of the archives has found no original source.

*I am very glad to hear that you are so profoundly interested in Spinoza's work. I too have a great admiration for the man but my knowledge of his works is not very profound.

Letter of November 23, 1951, offered on eBay October 16, 2008. Addressee is not yet identified.

On Adlai Stevenson (1900–1965)

Stevenson is very talented, but he doesn't make good use of it.

On the former Democratic Illinois governor who unsuccessfully ran for president against Eisenhower in 1952 and 1956. Quoted by Fantova, "Conversations with Einstein," December 12, 1953

On Rabindranath Tagore (1861–1941)

The verbal dialogue with Tagore was a complete disaster because of difficulties in communication, and should never have been published.

> To Romain Rolland, October 10, 1930, on the conversation with Indian mystic, poet, musician, and Nobel laureate in literature (1913) that took place in summer 1930, published in the *New York Times Magazine* on August 10, 1930. (A second discussion took place August 19.) Though journalist Dmitri Marianoff claimed Einstein personally vetted the article before publication, Einstein now appears to have changed his mind on its accuracy or in fact did not have a chance to see it first. Einstein shortened Tagore's first name and nicknamed him "Rabbi Tagore." Einstein Archives 33-029

On Nikola Tesla (1856–1943)

*I am happy that, as you celebrate your 75th birthday, you, as an eminent pioneer in the realm of high-frequency currents, have been able to witness the wonderful technical developments enabled by this field. I congratulate you on the great success of your life's work.

> To Tesla, June 1931. Einstein Archives 48-566

On Leo Tolstoy (1828–1910)

I doubt if there has been a true moral leader of worldwide influence since Tolstoy. . . . He remains in many ways the foremost prophet of our time. . . . There is no one today with Tolstoy's deep insight and moral force.

On the great Russian novelist. From an interview, *Survey Graphic* 24 (August 1935), 384, 413

On Arturo Toscanini (1867–1957)

You are not only the unrivaled interpreter of world musical literature, whose construction deserves the highest admiration, [but] in your fight against the fascist miscreants you have also shown yourself to be a highly conscientious man.

To the Italian conductor, composer, and pianist, March 1, 1936. Einstein Archives 34-386

Only one who devotes himself to a cause with his whole strength and soul can be a true master. For this reason, mastery demands the whole person. Toscanini demonstrates this in every manifestation of his life.

From a statement about Arturo Toscanini on the presentation of the American Hebrew Medal to him in 1938. The musician was strongly opposed to German and Italian fascism and left Europe for the United States in the mid-1930s. Einstein Archives 34-390

On Raoul Wallenberg (1912–?)

As an old Jew, I appeal to you to do everything possible to find and send back to his country Raoul Wallenberg, who was one of the very few who, during the bad years of the Nazi persecution, on his own accord and risking his own life, worked to rescue thousands of my unhappy Jewish people.

> To Joseph Stalin, November 17, 1947. An underling replied that a search proved unsuccessful. Wallenberg, a Swede, is credited with saving the lives of tens of thousands of Hungarian Jews during World War II. He was eventually captured, not by the Nazis but by the Soviet Army when it entered Budapest in 1945, and was never seen again. His fate is unknown. See Roboz Einstein, *Hans Albert Einstein*, 14. Einstein Archives 34-750

On Chaim Weizmann (1874–1952)

The chosen one of the chosen people.

> To Chaim Weizmann, October 27, 1923. In 1949, he became the first president of Israel. Einstein Archives 33-366

My feelings toward Weizmann are ambivalent, as Freud would say.

> Said to Abraham Pais, 1947. See Pais, *A Tale of Two Continents*, 228

On Hermann Weyl (1885–1955)

He is a very remarkable mind but a little removed from reality. In the new edition of his book, he made a complete mess of relativity, I think—God forgive him. Perhaps he will eventually realize that, for all his keen perceptions, he has shot wide off the mark.

To Heinrich Zangger, February 27, 1920, on the German-Swiss-American physicist who later joined the Institute for Advanced Study faculty the same year as Einstein. *CPAE*, Vol. 9, Doc. 332

On John Wheeler (1911–2008)

What Wheeler told me left a big impression on me, but I don't think I'll live to find out who is correct. . . . It was the first time I heard something sensible. . . . A possibility would be a combination of his ideas and mine.

Quoted by Fantova, "Conversations with Einstein," November 11, 1953, about the Princeton University theoretical physicist who in 1967 would coin the term "black hole"

On Woodrow Wilson (1856–1924)

*At first glance, his greatest contribution, the League of Nations, appears to have failed. Still, despite the fact that the League was crippled by his contemporaries and rejected by his own country, I have no doubt that Wilson's work will one day emerge in more effective form.

> On the twenty-eighth U.S. president (served 1913–1921) and former president of Princeton University (1902–1910), after which he became governor of New Jersey. From "I Am an American," June 22, 1940. See Rowe and Schulmann, *Einstein on Politics*, 471. Einstein Archives 29-092

Among the most important American statesmen it is probably Wilson who most represents the intellectual type. He, too, seemed not to have been very talented when it came to people skills.

> Ibid.

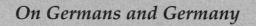

On Germans and Germany

*What you have said about German professors is not at all exaggerated. . . . Blind obedience to authority is the greatest enemy of the truth.

To Swiss teacher Jost Winteler, with whom he boarded while attending school in Aarau, complaining about a professor who would accept no criticism, July 8, 1901. *CPAE*, Vol. 1, Doc. 115

These are not really people with natural sentiments; they are cold and of an odd mixture of class-based condescension and servility, without showing goodwill toward their fellow men. Ostentatious luxury alongside creeping misery in the streets.

To Michele Besso, May 13, 1911, commenting on the Germans of Prague. *CPAE*, Vol. 5, Doc. 267

Now I understand the complacency of the citizens of Berlin. One gets so much outside stimulation here that one doesn't feel one's own emptiness as profoundly as one would in a more tranquil little spot.

To the Hurwitz family, May 4, 1914. *CPAE*, Vol. 8, Doc. 6

It is extraordinarily inspiring here in Berlin.

To Wilhelm Wien, June 15, 1914. *CPAE*, Vol. 8, Doc. 14

The people are like anywhere else, but you can make a better selection because there are so many of them.

To Robert Heller, July 20, 1914, on Berlin. *CPAE*, Vol. 8, Doc. 25

*I know people in Germany whose private lives are guided by virtually unbounded altruism, but who were awaiting the declaration of unlimited submarine warfare with the greatest impatience. . . . These people must be shown that it is necessary to have consideration for non-Germans as worthy equals, that it is essential to earn the *trust* of foreign countries, in order to be able to exist, that the goals that one sets for oneself cannot be achieved through force and treachery.

To Romain Rolland, August 22, 1917. *CPAE*, Vol. 8, Doc. 374

*Germans who love culture will soon again be able to be as proud of their fatherland as ever—and with more justification than before 1914.

To Arnold Sommerfeld, December 6, 1918. *CPAE*, Vol. 8, Doc. 665

The country is like a man with a badly upset stomach who has not yet vomited enough.

To Aurel Stodola, March 31, 1919, on the right-wing putsch in Berlin two weeks earlier. *CPAE*, Vol. 9, Doc. 16

People here appeal to me better in misfortune than in fortune and plenty.

To Heinrich Zangger, June 1, 1919. *CPAE*, Vol. 9, Doc. 52

*These people have no clear notion that they have become the blind instruments of an arrogant and unscrupulous minority, which is why their professed indignation about a "dictated peace" is not a hollow phrase or sham, but reflects their actual experience.

To Adriaan Fokker, July 30, 1919. *CPAE*, Vol. 9, Doc. 78

Ever since things have been going badly for the people here, they appeal to me incomparably better. Misfortune suits mankind immeasurably better than success.

To Heinrich Zangger, December 24, 1919. *CPAE*, Vol. 9, Doc. 233

Berlin is the place to which I am most closely bound by human and scientific ties.

To K. Haenisch, Prussian Minister of Education, September 8, 1920. Einstein Archives 36-022

*There is here in Germany an undeniable irritability toward me because of my pacifist and other political attitudes, which is amplified by the country's difficult political situation.

To Evelyn Wagner, January 31, 1921. *CPAE*, Vol. 12, Doc. 38

Germany had the misfortune of becoming poisoned, first because of plenty, and then because of want.

Aphorism, 1923. Einstein Archives 36-591

Funny people, these Germans. To them I am a stink-
ing flower, yet they make me into a boutonniere
time and time again.

From his Travel Diary, April 17, 1925

Hitler is living—or shall I say sitting—on the empty
stomach of Germany. As soon as economic conditions improve, Hitler will sink into oblivion.

In *Cosmic Religion* (1931), 106–107. Perhaps a paraphrase.
Original source unknown.

As long as I have any choice, I will only stay in a
country where political liberty, tolerance, and equality of all citizens before the law prevail. . . . These
conditions do not obtain in Germany at the present
time.

From "Political Manifesto," a statement to the American
press before returning to Europe regarding the Hitler
regime, March 11, 1933. Reprinted in Rowe and Schulmann, *Einstein on Politics*, 269–270. Einstein Archives
28-235

The overemphasized military mentality in the German state was alien to me even as a boy. When my
father moved to Italy, he took steps, at my request,
to have me released from German citizenship because I wanted to become a Swiss citizen.

To Julius Marx, April 3, 1933. Quoted in Hoffmann,
Albert Einstein: Creator and Rebel, 26. Einstein Archives
51-070

The statements I have issued to the press concerned my intention to resign my position in the Academy and renounce my Prussian citizenship. I gave as my reason for these steps that I did not wish to live in a country where the individual does not enjoy equality before the law and freedom to say and teach what he likes.

To the Prussian Academy of Sciences, April 5, 1933. Einstein Archives 29-295

*I cannot help but remind you that, in all these years, I have only enhanced Germany's prestige and never allowed myself to be alienated by the systematic attacks on me in the rightist press, especially those of recent years in which no one took the trouble to stand up for me. . . . Is not the destruction of German Jews by starvation the official program of the present German government?

To Max Planck, April 6, 1933, after resigning from the Prussian Academy of Sciences. Einstein Archives 19-391

You have also remarked that a "good word" on my part for "the German people" would have produced a great effect abroad. To this I must reply that such testimony as you suggest would have been equivalent to a repudiation of all the notions of justice and liberty for which I have stood all of my life. Such testimony would not be, as you put it, a good word for the German nation.

From a reply to the Prussian Academy of Sciences, April 12, 1933, after it accepted Einstein's resignation. Einstein Archives 29-297

I have now been promoted to an "evil monster" in Germany, and all my money has been taken away from me. But I console myself with the thought that the latter would soon be gone, anyway.

To Max Born, May 30, 1933, after Einstein's German bank account had been confiscated. In Born, *Born-Einstein Letters*, 112. Einstein Archives 8-192

I cannot understand the passive response of the whole civilized world to this modern barbarism. Doesn't the world see that Hitler is aiming for war?

Quoted by a reporter for *Bunte Woche* (Vienna), October 1, 1933; also quoted in Pais, *Einstein Lived Here*, 194

Germany the way it used to be was [a cultural] oasis in the desert.

To Alfred Kerr, July 1934. Einstein Archives 50-687

Germany is still war-minded and conflict is inevitable. The nation has been on the decline mentally and morally since 1870. Many of the men I associated with in the Prussian Academy have not been of the highest caliber in the nationalistic years since the world war.

From an interview, *Survey Graphic* 24 (August 1935), 384, 413

You have found the right words for the great problems of our time, and they do not remain without effect. I only wish you would not have used the word "Aryans" as if it were a rational notion.

To Rabbi Stephen Wise, September 11, 1935, commenting on a speech Wise gave in Lucerne. Einstein Archives 35-172

*It is well known that German Fascism has been particularly violent in its attack upon my Jewish brothers. . . . The alleged reason for this persecution is the desire to purify the "Aryan" race in Germany. As a matter of fact, no such "Aryan" race exists; this fiction has been invented solely to justify the persecution and expropriation of the Jews.

From a CBS radio broadcast sponsored by the American Christian Committee for German Refugees and the Emergency Committee in Aid of Polish Refugees from Nazism, October 22, 1935. See Rowe and Schulmann, *Einstein on Politics*, 293. Einstein Archives 28-317

*For centuries the German people have been subject to indoctrination by an unending succession of schoolmasters and drill sergeants. The Germans have been trained in hard work and were made to learn many things, but they have also been drilled in slavish submission, military routine and brutality. The postwar democratic Constitution of the Weimar Republic fitted the German people about as well as the giant's clothes fitted Tom Thumb.

From an unpublished manuscript, 1935; published later
in compilations: Nathan and Norden, *Einstein on Peace*,
263–264; Dukas and Hoffman, *Albert Einstein, the Human
Side*, 110; Rowe and Schulmann, *Einstein on Politics*, 295.
Einstein Archives 28-322

The only good thing I can see in this is that Hitler,
obsessed with his power, will engage in enough stupidities to unite the world against Germany—an
ever more grotesque version of Kaiser Wilhelm.

To Otto Nathan, September 15, 1936. Bergreen Albert Einstein Collection, Vassar College, Box M2003-009, Folder 1.15.
Einstein Archives 38-507

*In 1919 the Academy urged me to accept German
citizenship in addition to my Swiss one. I was stupid enough to give in.

To Mileva Einstein-Marić, July 20, 1938. Translated in
Neffe, *Einstein*, 276. Einstein Archives 75-949

They have always had the tendency to serve psy-
chopaths slavishly. But they have never been able to
accomplish it so successfully as at the present time.

Written in August as a scribble on the reverse of a letter
dated July 28, 1939, alluding to Hitler. Einstein Archives
28-500.1

Because of their wretched traditions, the Germans
are so evil that it will be very difficult to remedy the
situation by sensible, not to say humane, means.

I hope that by the end of the war they will largely kill themselves off with the kindly help of God.

> To Otto Juliusburger, Summer 1942. Quoted in Sayen, *Einstein in America*, 146. Einstein Archives 38-199

The Germans as an entire people are responsible for these mass murders and must be punished as a people. . . . Behind the Nazi Party stand the German people, who elected Hitler after he had in his book and in his speeches made his shameful intentions clear beyond the possibility of misunderstanding.

> On the heroes of the Warsaw Ghetto, in *Bulletin of the Society of Polish Jews* (New York), 1944. Einstein Archives 29-099

Since the Germans massacred my Jewish brethren in Europe, I will have nothing further to do with Germans, including a relatively harmless academy. This does not include those few who remained level-headed within the range of possibility.

> To Arnold Sommerfeld, December 14, 1946. Einstein included Otto Hahn, Max von Laue, Max Planck, and Arnold Sommerfeld among the few. Einstein Archives 21-368

The crime of the Germans is truly the most abominable ever to be recorded in the history of the so-called civilized nations. The conduct of the German intellectuals—seen as a group—was no better than that of the mob.

> To Otto Hahn, January 26, 1949. Einstein Archives 12-072

The attitude of the overwhelming majority of the Germans toward our people was such that we cannot help consider them anything but a danger. I judge the relations of the Germans to other nations to be equally dangerous.

> From an interview with Alfred Werner, *Liberal Judaism* 16 (April–May 1949), 4–12

After the mass murders that the Germans committed against the Jewish people it should be evident that a self-respecting Jew does not want to be associated with any official German event. My membership renewal in the Orden pour le mérite is therefore out of the question.

> To German president Theodor Heuss, January 16, 1951. Einstein was given this prestigious medal in 1923, but was forced by the Nazis to return it in 1933. He was invited by the German government to rejoin the order in 1950, and as the above letter shows, he declined. Charles Darwin, Michael Faraday, Max Planck, Walther Nernst, David Hilbert, the Grimm Brothers, and Richard Strauss were among the hundreds of artists and scientists who had received the medal since 1842. Einstein Archives 34-427

If I were wrong, then one [author] would have been enough!

> Einstein's retort with regard to his theory when he heard that a book titled *100 Authors against Einstein* was published in Germany. Quoted in Stephen Hawking, *A Brief History of Time* (London: Bantam, 1988), 178

On Humankind

At times such as this, one realizes what a sorry species one belongs to. I am moving along quietly with my contemplations while experiencing a mixture of pity and revulsion.

> To Paul Ehrenfest, August 19, 1914, at the onset of World War I. *CPAE*, Vol. 8, Doc. 34

I see that often the most power-hungry and politically extreme people could not as much as kill a fly in their personal lives.

> To Paul Ehrenfest, June 3, 1917. *CPAE*, Vol. 8, Doc. 350

How is it possible that this culture-loving era could be so monstrously amoral? More and more I come to value charity and love of one's fellow being above everything else.

> To Heinrich Zangger, December 6, 1917. *CPAE*, Vol. 8, Doc. 403

Man tries to fashion for himself a simplified and intelligible picture of the world; he then tries to substitute this cosmos for his own world of experience. . . . Each makes this cosmos and its construction the pivot of his emotional life in order to find the peace and security he can't find in the narrow whirlpool of personal experience.

> From "Motives for Research," a lecture delivered at Max Planck's sixtieth birthday, April 1918. *CPAE*, Vol. 7, Doc. 7

*I am convinced that the effort of striving for an improved existence as a wage earner combined with the disdain people have toward those who are not gainfully employed are sufficiently strong psychological forces to ensure the healthy development of economic life.

> To the society "A Guaranteed Subsistence for All," December 12, 1918. *CPAE*, Vol. 7, Doc. 16

*God knows, I prefer people with anxieties, whose tomorrow is threatened by uncertainty.

> To Max Born, January 19, 1919, on a postcard from Switzerland. *CPAE*, Vol. 9, Doc. 3

Failure and deprivation are the best educators and purifiers.

> To Auguste Hochberger, July 30, 1919. *CPAE*, Vol. 9, Doc. 79

One should keep in mind that on average the moral qualities of people do not differ much from country to country.

> To H. A. Lorentz, August 1, 1919. *CPAE*, Vol. 9, Doc. 80

*Misfortune suits mankind immeasurably better than success.

> To Heinrich Zangger, December 24, 1919. *CPAE*, Vol. 9, Doc. 233

*It is not enough for us to play a part as individuals in the cultural development of the human race; we must also attempt tasks which only nations as a whole can perform.

From an address delivered June 27, 1921, to a Zionist meeting in Berlin, published in *Mein Weltbild* (1934). See *Ideas and Opinions*, 182. Einstein Archives 28-010

Children don't heed the life experiences of their parents, and nations ignore history. Bad lessons always have to be learned anew.

Aphorism, October 12, 1923. Einstein Archives 36-589

Why do people speak of great men in terms of nationality? Great Germans, great Englishmen? Goethe always protested against being called a German poet. Great men are simply men and are not to be considered from the point of view of nationality, nor should the environment in which they were brought up be taken into account.

Quoted in the *New York Times*, April 18, 1926, 12:4

*Reading, after a certain age, diverts the mind too much from its creative pursuits. Any man who reads too much and uses his own brain too little falls into lazy habits of thinking, just as the man who spends too much time in the theater is tempted to be content with living vicariously instead of living his own life.

From an interview with G. S. Viereck, "What Life Means to Einstein," *Saturday Evening Post*, October 26, 1929; reprinted in Viereck, *Glimpses of the Great*, 437

I look upon mankind as a tree with many sprouts. It does not seem to me that every sprout and every branch possesses an individual soul.

Ibid., 444

*People are like bicycles. They can keep their balance only as long as they keep moving.

To son Eduard, February 5, 1930. Translated in Neffe, *Einstein*, 369. Einstein Archives 75-590

It is people who make me seasick—not the sea. But I am afraid that science has yet to find a solution for this ailment.

To the Schering-Kahlbaum company in Berlin, after it sent him some anti-seasickness samples. November 28, 1930. Einstein Archives 48-663

Enjoying the joys of others and suffering with them—these are the best guides for men.

To Valentine Bulgakov, November 4, 1931. Einstein Archives 45-702

*Doing "good" (in an ethical sense) is in reality difficult to achieve. To me, something creative that has been done with love is to have done "good."

Ibid.

Before God we are relatively all equally wise—
equally foolish.

In *Cosmic Religion* (1931), 105. Original source unknown.

The true value of a human being is determined primarily by the measure and the sense in which he has
attained liberation from the self.

Ca. June 1932. Reprinted in *Mein Weltbild*, 10; and *Ideas and
Opinions*, 12

The minority, presently the ruling class, has the
school and the press, and usually the Church as
well, under its thumb. This enables it to organize
and sway the emotions of the masses, and use them
as its tool.

To Sigmund Freud, July 30, 1932. In *Why War?* 5. Einstein
Archive 32-543

In the last analysis, everyone is a human being,
whether he is an American or a German, a Jew or a
Gentile. If it were possible to hold only this worthy
point of view, I would be a happy man.

To Gerald Donahue, April 3, 1935. Einstein Archives
49-502

Man owes his strength in the struggle for existence to
the fact that he is a socially living animal. As little as
a battle between single ants of an ant hill is essential

for survival, just so little is this the case with individual members of a human community.

> From an address at the celebration of the tercentenary of
> higher education in America at the State University of New
> York in Albany, October 15, 1936. In *School and Society* 44
> (1936), 589–592. Published as "On Education," in *Ideas and
> Opinions*, 62. Einstein Archives 29-080

A successful man is one who receives a great deal
from his fellow-men, usually incomparably more
than corresponds to *his* service to them. The value of
a man, however, should be seen in what he gives
and not in what he is able to receive.

> Ibid.

*All religions, arts, and sciences are branches of the
same tree. All these aspirations are directed toward
ennobling man's life, lifting it from the sphere of
mere physical existence and leading the individual
toward freedom.

> Opening words to "Moral Decay," a message to Young
> Men's Christian Association on its Founder's Day, Octo-
> ber 11, 1937. Quoted in *Out of My Later Years*, 16. Einstein
> Archives 28-403

One misses the elementary reaction against injustice
and for justice—that reaction which in the long run
represents man's only protection against a relapse
into barbarism.

> From ibid.

When one looks at humankind today, one notices with regret that quantity does not make up for quality: if quantity could only substitute for quality, we would be in better circumstances now than was Ancient Greece.

To Ruth Norden, December 21, 1937. Einstein Archives 86-933

Common convictions and aims, similar interests, will in every society produce groups that, in a certain sense, act as units. There will always be friction between such groups—the same sort of aversion and rivalry that exists between individuals. . . . In my opinion, uniformity in a population would not be desirable, even if it were attainable.

From "Why Do They Hate the Jews?" *Collier's* magazine, November 26, 1938

*And yet I know that, all in all, man changes but little, even though prevailing notions make him appear in a very different light at different times, and even though current trends like the present bring him unimaginable sorrow. Nothing of all that will remain but a few pitiful pages in the history books, briefly picturing to the youth of future generations the follies of its ancestors.

From "Ten Fateful Years," written for Clifton Fadiman's book, *I Believe* (1939). See Rowe and Schulmann, *Einstein on Politics*, 312–314, for full text.

The ancients knew something that we seem to have forgotten: that all means are but a blunt instrument if they do not have a living spirit behind them.

> From an address at Princeton Theological Seminary, May 19, 1939. Einstein Archives 28-493

It is better for people to be like the beasts . . . they should be more intuitive; they should not be too conscious of what they are doing while they are doing it.

> From a conversation recorded by Algernon Black, Fall 1940. Einstein forbade the publication of this conversation. Einstein Archives 54-834

We have to do the best we are capable of. This is our sacred human responsibility.

> Ibid.

*The satisfaction of physical needs is indeed the indispensable precondition of a satisfactory existence, but in itself it is not enough. In order to be content, men must also have the possibility of developing their intellectual and artistic powers to whatever extent accords with their personal characteristics and abilities.

> From "On Freedom," ca. 1940, excerpted from "On Freedom and Science," in Ruth Anshen, *Freedom: Its Meaning* (Harcourt, Brace, 1940). See Rowe and Schulmann, *Einstein on Politics*, 433

*Only if outward and inner freedom are constantly and consciously pursued is there a possibility of spiritual development and perfection and thus of improving man's outward and inner life.

Ibid., 435

Perfection of means and confusion of goals seem, in my opinion, to characterize our age.

From a broadcast recording for a science conference in London, September 28, 1941. Einstein Archives 28-557

*I must confess that biographies have seldom attracted me or captured my imagination. Autobiographies mostly arise out of narcissism or negative feelings toward others. Biographies from the pen of another person tend in their psychological traits to reflect the intellectual and spiritual nature of the writer more than that of the person portrayed.

From the foreword to Philipp Frank, *Einstein*, and *Albert Einstein: Sein Leben und Seine Zeit*. Foreword written ca. 1942. Einstein Archives 28-581

*The striving for truth and knowledge is one of the highest of man's qualities—though often the pride is most loudly voiced by those who strive the least.

From an NBC radio broadcast, "The Goal of Human Existence," for the United Jewish Appeal, April 11, 1943. See Rowe and Schulmann, *Einstein on Politics*, 322. Einstein Archives 28-587

*The feeling of what ought and ought not to be grows and dies like a tree, and no fertilizer of any kind will do very much good. The individual can only give a fine example and have the courage to uphold ethical convictions sternly in a society of cynics.

To Max Born, September 7, 1944. In *Born-Einstein Letters*,
145. Einstein Archives 8-207

*We humans generally live under the illusion of security and being at home and hearth in a trusting and gracious physical and human environment. When the daily rhythm is interrupted, however, we realize we are like shipwrecked people trying to balance ourselves on a wretched plank in the open sea, forgetting where we came from and not knowing how to steer.

To Mr. and Mrs. Held, who had lost a child or grandchild,
April 26, 1945. Einstein Archives 56-853

*One should strive to attain [individual freedom] to the degree that this is compatible with the need to protect the security of the individual and to provide essential economic needs. In other words: the first priority is a secure life, and then comes fulfillment of the need for freedom.

From "Is There Room for Individual Freedom in a Socialist
State?" ca. July 1945. See Rowe and Schulmann, *Einstein on
Politics*, 436. Einstein Archives 28-661

*He who is untrue to his own cause cannot command the respect of others.

From a statement in *Aufbau* 11, no. 50 (December 14, 1945), 11

There is no greater satisfaction for a just and well-meaning person than the knowledge that he has devoted his best energies to the service of a good cause.

Closing words to "A Message to My Adopted Country," *Pageant* 1, no. 12 (January 1946), 36–37. See Rowe and Schulmann, *Einstein on Politics*, 476; Jerome and Taylor, *Einstein on Race and Racism*, 141

*A large part of our attitude toward things is conditioned by opinions and emotions which we unconsciously absorb as children from our environment. In other words, it is tradition—besides inherited aptitudes and qualities—which makes us what we are. . . . We must try to recognize what in our accepted tradition is damaging to our fate and dignity—and shape our lives accordingly.

Ibid.

*Man is born to hate in almost higher measure than to love, and hate does not tire in taking grasp of any available situation.

To Hans Muehsam, April 3, 1946. Einstein Archives 38-352

I think we have to safeguard ourselves against people who are a menace to others, quite apart from
what may have motivated their deeds.

To Otto Juliusburger, April 11, 1946. Here Einstein was referring specifically to Hitler. Einstein Archives 38-228

Perhaps there is something benevolent in the waste-
ful sport that Nature, seemingly blindly, places on
her creatures. It can only be good to try to persuade
young people how critical that decision [marriage
and reproduction] is—often made at the moment
when Nature leaves us in a kind of drunken sensual
delusion so that we least have our good judgment
when we most need it.

To Hans Muehsam, June 4, 1946, on Einstein's rejection of
any concerted effort to "improve" the human race. Einstein
Archives 38-356

*To act intelligently in human affairs is only possible
if an attempt is made to understand the thoughts,
motives, and apprehensions of one's opponent so
fully that one can see the world through his eyes.
All well-meaning people should try to contribute
as much as possible to improving such mutual
understanding.

"A Reply to the Soviet Scientists," December 1947, published in *Bulletin of the Atomic Scientists* 4, no. 2 (February
1948), 35–37. See also Rowe and Schulmann, *Einstein on
Politics*, 393

Man is, at one and the same time, a solitary being and a social being. As a solitary being, he attempts to protect his own existence and that of those who are closest to him, to satisfy his personal desires, and to develop his innate abilities. As a social being, he seeks to gain the recognition and affection of his fellow human beings, to share in their pleasures, to comfort them in their sorrows, and to improve their conditions of life.

From "Why Socialism?" *Monthly Review*, May 1949. See Rowe and Schulmann, *Einstein on Politics*, 441. Einstein Archives 28-857

*Man can find meaning in life, short and perilous as it is, only through devoting himself to society.

Ibid., 443

One is born into a herd of buffalos and must be glad if one is not trampled underfoot before one's time.

To Cornel Lanczos, July 9, 1952. Einstein Archives 15-320

[A man] must learn to understand the motives of human beings, their illusions, and their sufferings in order to acquire a proper relationship to individual fellow-men and to the community.

From an interview with Benjamin Fine, *New York Times*, October 5, 1952. Reprinted as "Education for Independent Thought" in *Ideas and Opinions*, 66. Einstein Archives 59-666

What a person thinks on his own, without being stimulated by the thoughts and experiences of other people, is even in the best case rather paltry and monotonous.

From an article in *Der Jungkaufmann* (Zurich) 27, no. 4 (1952), 73. Einstein Archives 28-927

You must be aware that most men (and also not only a few women) are by nature not monogamous. This nature makes itself even more forceful when tradition and circumstance stand in an individual's way.

To Eugenie Anderman, an unknown woman who was seeking advice from Einstein regarding her husband's infidelity, June 2, 1953. (Courtesy of Andor Carius.) Einstein Archives 59-097

A forced faithfulness is a bitter fruit for all concerned.

Ibid.

We all are nourished and housed by the work of our fellowmen and we have to pay honestly for it not only by the work we have chosen for the sake of our inner satisfaction but by work which, according to general opinion, serves them. Otherwise we become parasites, however modest our needs may be.

To a man who wanted to spend his time being subsidized to study rather than work, July 28, 1953. Einstein Archives 59-180

*A large part of history is . . . replete with the struggle for human rights, an eternal struggle in which a final victory can never be won. But to tire in that struggle would mean the ruin of society.

> From "Human Rights," a message to the Chicago Decalogue Society of Lawyers upon receiving its award for contributions to human rights. The message had been written just before December 5, 1953 (Einstein Archives 28-1012), and was translated and recorded before being played at the ceremony on February 20, 1954. See Rowe and Schulmann, *Einstein on Politics*, 497

The existence and validity of human rights are not written in the stars.

> Ibid.

To obtain an assured favorable response from people, it is better to offer them something for their stomachs instead of their brains.

> To L. Manners, a chocolate manufacturer, March 19, 1954. Einstein Archives 60-401

Fear or stupidity has always been the basis of most human actions.

> To E. Mulder, April 1954. Einstein Archives 60-609

*In matters concerning truth and justice there can be no distinction between big problems and small; for the general principles which determine the conduct of men are indivisible. Whoever is careless with the

truth in small matters cannot be trusted in important affairs.

> From an undelivered message to the world, April 1955, in-
> tended to address the Arab-Israeli conflict. Einstein died
> before he could finish and deliver the speech. See Rowe
> and Schulmann, *Einstein on Politics*, 506. Einstein Archives
> 28-1098.

The individual, if left alone from birth, would remain primitive and beastlike in his thoughts and
feelings. . . . The individual is what he is . . . by virtue [of being] a member of a great human community that directs his material and spiritual existence
from cradle to grave.

> From "Society and Personality," in *Ideas and Opinions*, 13

Whoever undertakes to set himself up as a judge of
Truth and Knowledge is shipwrecked by the laughter of the gods.

> Aphorism in *Essays Presented to Leo Baeck on the Occasion
> of His Eightieth Birthday* (London: East and West Library,
> 1954), 26. Leo Baeck was a rabbi and philosopher who led
> the Jewish community in Germany during the time of Hit-
> ler. See also the poem "The Wisdom of Dialectical Material-
> ism" in this volume. Rowe and Schulmann, *Einstein on Poli-
> tics*, 457, translate the aphorism a bit differently in their
> note. Einstein Archives 28-962

In order to be a perfect member of a flock of sheep,
one has to be, foremost, a sheep.

> Ibid.

Hail to the man who went through life always helping others, knowing no fear, and to whom aggres-
siveness and resentment are alien. Such is the stuff
of which the great moral leaders are made.

Ibid.

The attempt to combine wisdom and power has
only rarely been successful, and then only for a short
while.

Ibid.

Few people are capable of expressing with equanimity opinions that differ from the prejudices of
their social environment. Most people are even incapable of forming such opinions.

Ibid.

On Jews, Israel, Judaism, and Zionism

In his youth, Einstein did not identify strongly with Jewish culture and religion. His parents were assimilated Jews in southern Germany who had distanced themselves from their Jewish roots and were more interested in the realities of entrepreneurship and making a good living. However, he did receive private religious instruction in Judaism at home and at first embraced it intensely, only to reject it decisively by the age of twelve as his interest in science became stronger. He then declared himself "without religious affiliation." He "rediscovered" his Jewish roots after moving to Berlin and becoming aware of the prejudicial treatment of eastern European Jews. This was coincident both with the advent of anti-Semitism and Zionism and with the confirmation of his general theory of relativity in 1919, which brought him worldwide fame. Einstein's Zionism was cultural rather than political. It emphasized the cultural and spiritual renewal of the Jewish people, as opposed to political Zionism, which focused on the establishment of a Jewish state. Still, he supported the creation of Israel as a refuge for Jews because he believed in the power of a community as a cohesive force and felt Jews should have a safe place to study and teach. He favored a two-state solution for Palestine, but realized there was no turning back after the creation of Israel, and late in his life he supported the new state even as he advocated fairness to the Arabs. However, he also supported his friend Erich Kahler in admonishing Arab landowners and politicians who "did nothing to improve the nature, the civilization, or the living standards of their countries" (*Princeton Herald*, April 14 and 28, 1944). See Schulmann, "Einstein Rediscovers Judaism";

Stachel, "Einstein's Jewish Identity"; and Jerome, *Einstein on Israel and Zionism*.

It goes against the grain to travel without necessity to a country in which my kinsmen have been persecuted so brutally.

To P. P. Lazarev, May 16, 1914, responding to an invitation from the Imperial Academy of Sciences in St. Petersburg to come to Russia to watch the 1914 total solar eclipse. *CPAE*, Vol. 8, Doc. 7

One doesn't know where to get one's pleasure from human affairs. I get most of my joy from the emergence of the Jewish state in Palestine. It does seem to me that our kinfolk are really more sympathetic (at least less brutal) than those horrid Europeans.

To Paul Ehrenfest, March 22, 1919. *CPAE*, Vol. 9, Doc. 10

I have great confidence in a positive development for a Jewish state and am glad that there will be a little patch of earth on which our brethren are not considered aliens.

To Paul Epstein, October 5, 1919. *CPAE*, Vol. 9, Doc. 122

One can be internationally minded without being indifferent to one's kinsmen.

Ibid.

I have warm sympathy for the affairs of the new colony in Palestine and especially for the yet-to-be-founded university. I shall gladly do everything in my power for it.

> To Hugo Bergmann, Zionist Organization London, November 5, 1919. *CPAE*, Vol. 9, Doc. 155

From 16–18 January I have to be in Basel, where a conference is being held to consult about the Hebrew University to be founded in Palestine. I believe that this undertaking deserves zealous collaboration.

> To Michele Besso, December 12, 1919. *CPAE*, Vol. 9, Doc. 207

*In my opinion, aversion to Jews is simply based upon the fact that Jews and non-Jews are different. It is the same feeling of aversion that is always found when two nationalities have to deal with each other. This aversion is a consequence of the *existence* of Jews, not of any particular qualities. . . . The feeling of aversion toward people of a foreign race with whom one has, more or less, to share daily life will emerge by necessity.

> From "Assimilation and Anti-Semitism," April 3, 1920. *CPAE*, Vol. 7, Doc. 34. See also Rowe and Schulmann, *Einstein on Politics*, 144

I am neither a German citizen, nor do I believe in anything that can be described as a "Jewish faith." But I am a Jew and glad to belong to the Jewish people, though I do not regard it in any way as chosen.

To Central Association of German Citizens of the Jewish
Faith, April 3 [5], 1920. According to Rowe and Schulmann,
Einstein on Politics, 146, the letter shows how Einstein dis-
liked the association's tactic of self-defense. Published in
Israelitisches Wochenblatt, September 24, 1920. *CPAE,* Vol. 7,
Doc. 37, and Vol. 9, Doc. 368

*Only when we have the courage to regard ourselves
as a nation, only when we respect ourselves, can we
win the respect of others. . . . Anti-Semitism . . . will
always be with us as long as Jews and non-Jews are
thrown together. It may be thanks to anti-Semitism
that we owe our survival as a race; that at any rate is
what I believe.

Ibid.

*As much as I feel myself to be a Jew, I stand aloof
from the traditional religious rites.

To Jewish Community of Berlin, December 22, 1920. *CPAE,*
Vol. 9, Doc. 238

*The word "Jewish" has two meanings: it has to do
with (1) nationality and descent; (2) religion. I am a
Jew in the first sense but not in the second.

To Jewish Community of Berlin, January 5, 1921. *CPAE,*
Vol. 12, Doc. 8

I am not at all eager to go to America but am doing
it only in the interests of the Zionists, who must
beg for dollars to build educational institutions in

Jerusalem and for whom I act as high priest and decoy. . . . But I do what I can to help those in my tribe who are treated so badly everywhere.

To Maurice Solovine, March 8, 1921. Published in *Letters to Solovine*, 41. *CPAE*, Vol. 12, Doc. 85

*Naturally they don't need me for my abilities but because of my name, whose luster they hope will attract quite a bit of success with the rich kinsmen of Dollar-land. In spite of my emphatic internationalism, I believe that I am always under an obligation insofar as it is in my power to advocate on behalf of my persecuted and morally oppressed kinsmen.

To Fritz Haber, March 9, 1921. The letter is in regard to the Zionist Organization's request that Einstein accompany some of its members on a fund-raising trip to America for the Hebrew University of Jerusalem. Rowe and Schulmann, *Einstein on Politics*, 148. *CPAE*, Vol. 12, Doc. 88

*I confidently believe that the Jews will be prevented by the smallness and the dependency of their Palestinian colony from becoming obsessed with power.

To Maurice Solovine, March 16, 1921. *CPAE*, Vol. 12, Doc. 100

I know of no public event that has given me such pleasure as the proposal to establish a Hebrew University in Jerusalem. The traditional respect for knowledge that Jews have maintained intact through many centuries of severe hardship has made it particularly painful for us to see so many talented sons of the Jewish people cut off from higher education.

From an interview, *New York Times*, April 3, 1921. See similar remarks with respect to Brandeis University, below.

*Your leader, Dr. Weizmann, has spoken, and he has spoken very well for us all. Follow him and you will do well. That is all I have to say.

The full extent of a speech at a reception in New York for the Zionist Mission to America, April 12, 1921. Quoted in the *New York Times*, April 13, 1921, 13. See Illy, *Albert Meets America*, 91

*[I hope] that you do *not* mean that the *teaching* and the *research* of our future professors at Jerusalem are to be bound by the orthodox Jewish laws or conceptions. . . . Any such restriction of the freedom of speech or thought would be intolerable (except, perhaps, in a frankly theological institute or department) and would defeat your own purpose—to further a *free and creative* synthesis of faith and reason.

To Solomon Rosenbloom, April 27, 1921, on the Hebrew University in Jerusalem. *CPAE*, Vol. 12, Doc. 127

*The Hebrew university in Palestine will become a new "holy place" to our people.

Chicago Herald and Examiner, May 8, 1921, part 2, 3. See also Illy, *Albert Meets America*, 156

*Is it conceivable that, in addition to the tragedy of Jewish science without a home, there could exist a Jewish national home without science? The traditional pride of the Jewish people in their learned men would never suffer such humiliation.

Ibid., 158

*Zionism really represents a new Jewish ideal, one that can give the Jewish people renewed joy in existence. . . . I am very glad I accepted Weizmann's invitation. Some places, however, exhibit an overly intense Jewish nationalism that threatens to degenerate into intolerance and narrow-mindedness; but hopefully this is just a childhood disease.

To Paul Ehrenfest, June 18, 1921, on Einstein's fund-raising tour of the United States with Chaim Weizmann on behalf of the Hebrew University. *CPAE*, Vol. 12, Doc. 152

Anti-Semitism in Germany also has consequences that, from a Jewish point of view, should be welcomed. I believe German Jewry owes its continued existence to anti-Semitism. . . . Without this distinction, the assimilation of Jews in Germany would happen quickly and unimpeded.

From "How I Became a Zionist," *Jüdische Rundschau*, June 21, 1921. Rowe and Schulmann, *Einstein on Politics*, 151. *CPAE*, Vol. 7, Doc. 57

As long as I lived [in Switzerland], I was not aware of my Jewishness. . . . This changed as soon as I took

up residence in Berlin. . . . I saw how anti-Semitic
surroundings prevented [Jews] from pursuing reg-
ular studies and how they struggled for a secure
existence.

Ibid.

I am not a Jew in the sense that I call for the preservation of the Jewish or any other nationality as an
end in itself. I rather see Jewish nationality as a fact
and I believe that every Jew must draw the consequences from this fact. I consider raising Jewish self-
esteem essential, also in the interest of a natural co-
existence with non-Jews. This was my major motive
for joining the Zionist movement. . . . But my Zionism does not preclude cosmopolitan views.

Ibid., 152

Zionism, to me, is not just a colonizing movement
directed toward Palestine. The Jewish nation is a living fact in Palestine as well as in the Diaspora, and
Jewish national feelings must be kept alive wherever Jews live. . . . I believe that every Jew has an
obligation toward his fellow Jews.

Ibid., 152–153

*By leading Jews back to Palestine and restoring a
healthy and normal economic existence, Zionism
represents a productive activity that enriches all of

society. . . . [It] strengthens Jewish dignity and self-esteem, which are critical for existence in the Diaspora [and] . . . creates a strong bond that gives Jews a sense of self. I have always found repulsive the undignified addiction to conformity of many of my peers.

Ibid., 153

*To my knowledge this isn't planned. But since classes will be conducted in Hebrew, and because of the nationalist tendencies of its founders, in practical terms the [Hebrew] university will become a Jewish institution.

In answer to the question if the Hebrew University will be open only to Jews. From an interview in *Vorwärts* (Berlin), morning edition, June 30, 1921. See also letter to Solomon Rosenblooom, April 27, 1921, above. *CPAE*, Vol. 12, Appendix F

*I believe that membership in a religious denomination is unimportant. But among us Jews the conversion to a *different* religion is a symbolic act that implies that one wants to dissociate oneself from one's group. But one can well be *undenominational* without being disloyal to one's people. I myself am nonobservant [*konfessionslos*], yet consider myself a loyal Jew.

To Emil Starkenstein, July 14, 1921. *CPAE*, Vol. 12, Doc. 181

Where dull-witted kinsmen were praying aloud, their faces turned to the wall, their bodies rocking to

and fro. A pathetic sight of men with a past having
no place in present times.

> On his visit to the Wailing Wall in Jerusalem, February 3,
> 1923, recorded in his travel diary. Einstein Archives 29-129
> to 29-131

The heart says yes, but the mind says no.

> From his Travel Diary, February 13, 1923, on the invitation
> to accept a position at the Hebrew University of Jerusalem.
> In ibid. Also quoted in Hoffmann, "Einstein and Zionism,"
> 241. Einstein Archives 29-129

It is of no use to try to convince others of our spiritual and intellectual equality by way of reason, since
the others' attitude does not come by way of their
brains. We must instead emancipate ourselves socially and satisfy our own social requirements.

> From "Anti-Semitism and Academic Youth," April 1923. Re-
> printed in *Ideas and Opinions*, 188. Einstein Archives 28-016

*On the whole, the country [Palestine] is not very fer-
tile. It will become a moral center, but will not be able
to absorb a large proportion of the Jewish people.

> To Maurice Solovine, May 20, 1923. Einstein often spoke of
> Palestine as a "moral center," in which he envisioned a
> place of study and research for Jews without the onus of
> anti-Semitism. Einstein Archives 21-189

*By recalling to memory a past filled with glory and
sorrow and by opening their eyes to a healthier, dig-

nified future, Zionism can teach them self-knowledge and instill courage. It restores the moral force, which allows them to live and act in dignity. It frees the soul from the unforgivable feeling of exaggerated modesty, which can only oppress and make them unproductive. Finally it reminds them that the centuries they have lived through in common sorrow enjoins upon them the duty of solidarity.

From "Mission," in *Jüdische Rundschau* 14 (February 17, 1925), 129. First published in *La Revue Juive* (Geneva), January 15, 1925. See also Rowe and Schulmann, *Einstein on Politics*, 164–165. In this essay, according to Rowe and Schulmann, Einstein conceives of Palestine as "a model for a worldwide community of purpose based on international solidarity."

*The existence of this moral homeland will, I hope, succeed in imparting more vitality to a people that does not deserve to die. . . . For this reason I believe that I am able to assert that Zionism, which appears to be a nationalistic movement, has, when it comes down to it, a significant role to play for all mankind.

Ibid.

A Jew who strives to impregnate his spirit with humanitarian ideals can declare himself a Zionist without contradiction. What one must be thankful for to Zionism is that it is the only movement that has given Jews a justified pride.

Ibid.

Our Jews are up to much down there [Jerusalem] and are fighting among one another as usual. I actually have quite a bit to do with all of this because—as you know—I have become a Jewish saint.

> To Michele Besso, December 25, 1925. Einstein Archives 7-356. The tongue-in-cheek "Jewish saint" reference also appears in a letter to sons Hans Albert and Eduard, November 24, 1923 (see above, "On Einstein Himself"). In a letter to Paul Ehrenfest of May 4, 1920, he joked that "the bones of the saint were requested to be present" at a conference in Halle later that month.

*In the past our Jewish people have remained bound together by our traditions. They have had to pay for the blessings of this bond through a cultural narrowness that has led to vast spiritual and worldly limitations. Can this harm be undone without endangering ethnic survival? I think the following is possible: the life of the individual and community can proceed according to tradition; thinking should be allowed without restraint except as limited by the human mind.

> Foreword, written in German, to a Yiddish book on relativity by Tuvia Shalit, *The Special Theory of Relativity: Einstein's System and Minkowski's "World"* (Berlin: Self-published, 1927). (Thanks to Eli Maor for sending me the German original.)

*All Jewish children [in Palestine] should be obligated to learn Arabic.

> To Hugo Bergmann, September 27, 1929. Einstein Archives 37-768

Zionism is a nationalism which does not strive for power but for dignity and recovery.

Ibid.

*The Jews are a community bound together by ties of blood and tradition, and not of religion only: the attitude of the rest of the world toward them is sufficient proof of this. When I came to Germany fifteen years ago I discovered for the first time that I was a Jew, and I owe this more to Gentiles than Jews.

To Willy Hellpach, a noted psychologist and liberal politician during the Weimar era, October 8, 1929. See also Rowe and Schulmann, *Einstein on Politics*, 170. Einstein Archives 46-656

If we did not have to live among intolerant, narrow-minded, and violent people, I would be the first to discard all nationalism in favor of a universal humanity.

Ibid.

I am a determinist. As such, I do not believe in Free Will. The Jews believe in Free Will. They believe that man shapes his own life. I reject that doctrine philosophically. In that respect I am not a Jew.

From an interview with G. S. Viereck, "What Life Means to Einstein," *Saturday Evening Post*, October 26, 1929; reprinted in Viereck, *Glimpses of the Great*, 441

Should we be unable to find a way to honest cooperation and honest pacts with the Arabs, then we have
learned absolutely nothing from our 2,000 years of
suffering and will deserve our fate.

> To Chaim Weizmann, November 25, 1929. Einstein Ar-
> chives 33-411

*One who, like myself, has cherished the conviction
that humanity of the future must be built on an intimate community of all nations, and that aggressive
nationalism must be conquered, can see a future for
Palestine only on the basis of peaceful cooperation
between the two peoples who are at home in the
country. For this reason I should have expected that
the great Arab people will show a truer appreciation
of the need which the Jews feel to rebuild their national home in the ancient seat of Judaism.

> Letter to the Palestinian Arab paper *Falastin*, December 20,
> 1929, published February 1, 1930. Einstein Archives 46-148

Jewry has proved throughout history that the intellect is the best weapon. . . . It is our duty as Jews
to put at the disposal of the world our several-
thousand-year-old sorrowful experience and, true
to the ethical traditions of our forefathers, become
soldiers in the fight for peace, united with the noblest elements in all cultural and religious circles.

> Quoted in Frank, *Einstein: His Life and Times*, 156. Frank
> attributes the passage to a speech given at a "Jewish

meeting" in Berlin in 1929, though there is no trace of it in the archives.

The Jewish religion is . . . a way of sublimating everyday existence. . . . It demands no act of faith—in the popular sense of the term—on the part of its members. And for that reason there has never been a conflict between our religious outlook and the world outlook of science.

From "Science and God: A Dialogue," *Forum and Century* 83 (June 1930), 373

*Embedded in the tradition of the Jewish people there is a love of justice and reason which must continue to work for the good of all nations now and in the future.

From "The Jewish Community," a speech delivered at the Savoy Hotel, attended by George Bernard Shaw, H. G. Wells, and Lord Rothschild, October 29, 1930. See *Ideas and Opinions*, 174. Einstein Archives 29-033

*By means of modern methods of reconstruction, Palestine affords ample room for both Jews and Arabs, who can live side by side in peace and harmony in a common country. I believe that the setbacks of last year must strengthen within us the recognition of our duty to improve, through patience and continued efforts, our relations with the Arab people and to convince them of the advantages Zionism creates for them.

In "Redoubling Efforts," *New York Times*, December 3, 1930. From an interview given in Berlin to the *Jüdische Rundschau* shortly before leaving for the United States. See also Rowe and Schulmann, *Einstein on Politics*, 186

*[The] unity of Jews the world over is in no wise a political unity and should never become such. It rests exclusively on a moral tradition. Out of this alone can the Jewish people maintain its creative powers, and on this alone it claims its basis for existence.

From "The Jewish Mission in Palestine," radio address organized by the American Student Zionist Federation, which agreed with Einstein's cultural and socialist views for a Jewish state, December 13, 1930. See Rowe and Schulmann, *Einstein on Politics*, 187. Einstein Archives 28-121

Judaism is not a creed: the Jewish God is simply a negation of superstition, an imaginary result of its elimination. It is also an attempt to base moral law on fear, a regrettable and discreditable attempt. Yet it seems to me that the strong moral tradition of the Jewish nation has to a large extent shaken itself free from this fear. It is clear also that "serving God" was equated with "serving the living." The best of the Jewish people, especially the Prophets and Jesus, contended tirelessly for this.

From "Is There a Jewish Point of View?" *Opinion*, September 26, 1932; reprinted in *Ideas and Opinions*, 186. Einstein Archives 28-197

Judaism seems to me to be concerned almost exclusively with the moral attitude in life and toward life. . . . The essence of that conception seems to me to lie in an affirmative attitude toward the life of all creation.

> Ibid., 185. As reflected in this and similar statements, Einstein rejected any racial or other biological sanctions for Judaism; rather, he saw it primarily as an attitude toward life. See Stachel, "Einstein's Jewish Identity," for a discussion about Einstein's rejection of any taint of race in his concept of Judaism.

I consider the formation of labor unions . . . necessary for Palestine. For the working class is not only the very soul of construction work, but it is the only truly effective bridge between the Jews and Arabs.

> To Irma Lindheim, president of the Jewish women's organization, Hadassah, February 2, 1933. Einstein Archives 50-990

The Jews resemble an uncondensable noble gas that can assume a substantial form of existence only by adhering to a firm object. This applies to me as well. But perhaps it is in this very chemical inertia that our ability to act and to persist lies.

> To Paul Ehrenfest, June 1933. Einstein Archives 10-260

The pursuit of knowledge for its own sake, an almost fanatical love of justice, and the desire for personal independence—these are the features of the

Jewish tradition which make me thank my lucky stars that I belong to it.

> From "Jewish Ideals," written for the French periodical *VU/Temoignages: Les Juifs* in August 1933; reprinted in *Mein Weltbild*, 89, and *Ideas and Opinions*, 185

*In these days of the persecution of the German Jews especially it is time to remind the western world that it owes to the Jewish people (a) its religion and therewith its most valuable moral ideals, and (b), to a large extent, the resurrection of the world of Greek thought.

> Written for the French periodical *Cahiers Juifs* in 1933; reprinted in *The World as I See It* under the title "A Foreword" and in *Mein Weltbild* as "Deutsche und Juden." Einstein Archives 28-242

*Today the Jews of Germany find their fairest consolation in the thought of all they have produced and achieved for humanity by their efforts in modern times as well; and no oppression however brutal, no campaign of calumny however subtle will blind those who have eyes to see the intellectual and moral qualities inherent in this people.

> Ibid.

*We all know that the Jewish people has sustained itself through 2,000 years of severe hardships because it has regarded a tradition of *love for the spiritual and moral* as its highest possession.

From the acceptance speech for an honorary degree at Yeshiva College, October 8, 1934. Quoted in *Yeshiva University
News* (online), October 31, 2005

*Palestine will be a center of culture for all Jews, a
refuge for the most grievously oppressed, a field of
action for the best among us, a unifying ideal, and a
means of attaining inward health for the Jews of the
whole world.

An appeal for "Keren Hajessod," in *Mein Weltbild* (1934),
102, and *Ideas and Opinions*, 184

There are no German Jews, there are no Russian
Jews, there are no American Jews. Their only differ-
ence is their daily language. There are in fact only
Jews.

From a speech at a Purim dinner at the German-Jewish
Club in New York, March 24, 1935, reprinted shortly
thereafter in the *New York Herald Tribune*. A reader took
issue with this statement and asked Einstein to explain
what he meant. See the next quotation for a partial
answer.

Seen through the eyes of the historian, their history
of suffering teaches us that the fact of being a Jew
has had a greater impact than the fact of belonging
to political communities. If, for example, the German Jews were driven from Germany, they would
cease to be Germans and would change their language and their political affiliation; but they would

remain Jews. . . . I see the reason not so much in racial characteristics as in firmly rooted traditions that
are by no means limited to the area of religion.

> To Gerald Donahue, April 3, 1935. Note, however, that Ein-
> stein himself continued to speak the German language
> even after he was driven from Germany. Einstein Archives
> 49-502

Materialistic shallowness is a far greater menace to
the survival of the Jew than the numerous external
foes who threaten his existence with violence.

> From a message for a testimonial dinner, June 7, 1936.
> Quoted in the *New York Times*, June 8, 1936. Einstein Ar-
> chives 28-357

*I should much rather see reasonable agreement with
the Arabs on the basis of living together in peace
than the creation of a Jewish state. Apart from practical consideration, my awareness of the essential nature of Judaism resists the idea of a Jewish state with
borders, an army, and a measure of temporal power
no matter how modest. I am afraid of the inner damage Judaism will sustain—especially from the development of a narrow nationalism within our own
ranks, against which we have already had to fight
strongly, even without a Jewish state. We are no longer the Jews of the Maccabee period.

> From a speech titled "Our Debt to Zionism," before the Na-
> tional Labor Committee for Palestine on April 17, 1938, in

New York. Full text published in *New Palestine* 28, no. 16 (April 29, 1938), 2–4. Also see Rowe and Schulmann, *Einstein on Politics*, 300–302. Einstein Archives 28-427

Judaism owes a great debt of gratitude to Zionism. The Zionist movement has revived among Jews a sense of community. It has performed productive work . . . in Palestine, to which self-sacrificing Jews throughout the world have contributed. . . . In particular, it has been possible to lead a not inconsiderable part of our youth toward a life of joyous and creative work.

Ibid., 300

To be a Jew, after all, means, first of all, to acknowledge and follow in practice those fundamentals in humaneness laid down in the Bible—fundamentals without which no sound and happy community can exist.

Ibid.

The Jews as a group may be powerless, but the sum of the achievements of their individual members is everywhere considerable and telling, even though those achievements were made in the face of obstacles.

From "Why Do They Hate the Jews?" *Collier's* magazine, November 26, 1938. Reprinted in *Ideas and Opinions*, 197

[The Nazis] see the Jews as a nonassimilable element that cannot be driven into uncritical acceptance and that . . . threatens their authority because of its insistence on popular enlightenment of the masses.

Ibid.

The Jew who abandons his faith (in the formal sense of the word) is in a position similar to a snail that abandons its shell. He remains a Jew.

Ibid.

The bond that has united the Jews for thousands of years and that unites them today is, above all, the democratic ideal of social justice, coupled with the ideal of mutual aid and tolerance among all men. . . . The second characteristic of Jewish tradition is the high regard in which it holds every form of intellectual aspiration and spiritual effort.

Ibid.

*In the past we were persecuted despite the fact that we were the people of the Bible; today, however, it is just because we are the people of the Book.

From a CBS radio address for the United Jewish Appeal, March 21, 1939. See also Jerome, *Einstein on Israel and Zionism*, 141. Einstein Archives 28-475

The power of resistance which has enabled the Jewish people to survive for thousands of years has been based to a large extent on traditions of mutual helpfulness. In these years of affliction our readiness to help one another is being put to an especially severe test. May we stand this test as well as did our fathers before us. We have no other means of self-defense than our solidarity and our knowledge that the cause for which we are suffering is a momentous and sacred cause.

To Alfred Hellman, June 10, 1939. Einstein Archives 53-391

*I dislike Nationalism very much—even Jewish Nationalism. But our own national solidarity is forced upon us by a hostile world and not by the aggressive feelings which we connect with the word.

To Judge Jerome Frank, November 19, 1945. See Jerome, *Einstein on Israel and Zionism,* 157. Einstein Archives 35-071

*The cause for the fact that we Jews are a "separate people" is not only that *we* have the desire to set oursel[ves] apart, but that we are treated and persecuted as a separate people.

Ibid.

*The Jewish people are united not only through a common religion but also through common dangers and common problems of a political and social nature.

To Rabbi Louis Wolsey, November 20, 1945. Einstein Archives 35-075

*The enterprise to settle 30,000 Jewish war orphans in Birobidjan and secure for them . . . a satisfying and happy future is new proof for the humane attitude of Russia towards our Jewish people.

From a statement on Birobidjan, an autonomous Jewish region in Russia, December 10, 1945. Cited in Jerome, *Einstein on Israel and Zionism*, 158. Einstein Archives 56-517

*I was never in favor of a state. The [Jewish] state idea is not according to my heart. I cannot understand why it is needed. It is connected with many difficulties and narrow-mindedness. I believe it is bad.

Einstein's answer to a question posed by the Anglo-American Committee of Inquiry in Washington, established to assess the feasibility of Jewish immigration to Palestine and how it would affect both Arabs and Jews, January 11, 1946. Einstein favored a trusteeship of several nations to administer Palestine, to be set up by the United Nations, rather than a political state. Much of the hearing is reprinted in Rowe and Schulmann, *Einstein on Politics*, 340–344. According to Rowe and Schulmann, Einstein accepted the Jewish state "as the inevitable consequence of Britain's failure to create a workable political settlement in which Arabs and Jews could live together in peace" (p. 38). A full transcription of the proceedings is in Jerome, *Einstein on Israel and Zionism*, 161–175

*The difficulties between the Jews and the Arabs are artificially created, and are created by the English.

Ibid. Rowe and Schulmann, *Einstein on Politics*, 340

*If we had power it might be worse still. We imitate the stupid nationalism and racial nonsense of the *goyim* even after having gone through a school of suffering without equal.

Ibid., 346

Zionism gave the German Jews no great protection against annihilation. But it did give the survivors the inner strength to endure the debacle with dignity and without losing their healthy self-respect.

To Charles Adler, an anti-Zionist Jew, probably January 1946. Quoted in Dukas and Hoffmann, *Albert Einstein, the Human Side*, 64. Einstein Archives 56-435

*Animals without wings take care not to change their habitat unless the need arises. It follows that well-established groups must seek out the best environment when nothing stands in their way.

To Hans Muehsam, April 3, 1946. Einstein Archives 38-352

Under existing conditions, our young scientific talents frequently have no access to scholarly professions, which means that our proudest tradition—the appreciation of productive work—would be faced with slow extinction if we remain as inactive in this area as in the past.

> To David Lilienthal, July 1946, on his approval of the founding of Brandeis University to serve Jewish students. Within a year he added that it should "not know of discrimination for or against anybody because of sex, color, creed, national origin, or political opinion." In 1953, he refused an honorary doctorate from the institution because of some personal disputes he had had with the founders in 1947. Einstein Archives 40-398, 40-432. See also S. S. Schweber, "Albert Einstein and the Founding of Brandeis University," unpublished manuscript.

*We advocated irresponsible and unjust gains in Palestine under the influence of demagogues and erstwhile blowhards. . . . We are imitating the *goyim* in their idiotic nationalism and racism.

> To Hans Muehsam, January 22, 1947. Einstein Archives 38-361

The plight of the surviving victims of German persecution bears witness to the degree to which the moral conscience of mankind has weakened. Today's meeting shows that not all men are prepared to accept the horror in silence.

> From a message on the dedication of the Riverside Drive Memorial, New York, to the victims of the Holocaust, October 19, 1947. Einstein Archives 28-777

*I also think that in these latter years an agreement between us and the Arabs that could have led to a two-state solution was no longer possible. In the past—essentially since 1918—we have ignored the Arabs and entrusted them repeatedly to the English. I have never held the idea of a [Jewish] state to be desirable, for civic, political, and military reasons. But now there is no going back and [the situation] must be wrangled with.

To Hans and Minna Muehsam, September 24, 1948. Einstein Archives 38-380

The wisdom and moderation the leaders of the new state have shown gives me confidence that gradually relations will be established with the Arab people which are based on fruitful cooperation and mutual respect and trust. For this is the only means through which both peoples can attain true independence from the outside world.

From a statement to the Hebrew University of Jerusalem upon receipt of an honorary doctorate, March 15, 1949. Einstein Archives 28-854, 37-296

This university is today a living thing, a home of free learning and teaching and happy collegial work. There it is, on the soil that our people have liberated under great hardships; there it is, a spiritual center of a flourishing and buoyant community whose accomplishments have finally met with the universal recognition they deserve.

Ibid.

Zionism in 1921 strove for the establishment of a national home, not the foundation of a state in the political sense. However, this latter aim has been realized because of the pressure of necessity rather than emergency. To discuss this development retrospectively seems to be academic. As far as the attitude commonly described as "orthodoxy" is concerned, I never had much sympathy for it. Nor do I think it plays an important role now, or that it is likely to in the future.

In answer to interviewer Alfred Werner's question about his support of Zionism in the establishment of Israel as a secular state, in *Liberal Judaism* 16 (April–May 1949), 4–12

I must be very careful not to do any foolish thing or to write any foolish book in order to live up to that distinction. I am proud of the honor, not on my own account, but because I am a Jew. It certainly denotes progress when a Christian church honors a Jewish scientist.

Referring to his sculptured image over the main entrance of Riverside Church in New York City, which also depicts other immortal leaders of humanity. Ibid. Riverside Church, modeled after the thirteenth-century cathedral in Chartres, France, was completed in 1930 and is interdenominational.

The Jews of Palestine did not fight for political independence for its own sake, but they fought to achieve free immigration for the Jews of many countries where their very existence was in danger; free immigration also for all those who were longing for a life among their own. It is no exaggeration to say that they fought to make possible a sacrifice that is perhaps unique in history.

> From an NBC radio broadcast for a United Jewish Appeal conference, Atlantic City, November 27, 1949. See Rowe and Schulmann, *Einstein on Politics*, 353. Einstein Archives 28-862

*It was much less our own fault or that of our neighbors than of the Mandatory Power, that we did not achieve an undivided Palestine in which Jews and Arabs would live as equals, free, in peace. If one nation dominates other nations, as was the case in the British Mandate over Palestine, she can hardly avoid following the notorious device of *divide et impera*.

> Ibid.

The support for cultural life is of primary concern to the Jewish people. We would not be in existence today as a people without this continued activity in learning.

> From a statement on the occasion of the twenty-fifth anniversary of the Hebrew University of Jerusalem. Quoted in the *New York Times*, May 11, 1950

Jewry is a group of people with a common history and with certain traditions besides the religious one. They are united by common interests created and maintained by the outside world through a mostly antagonistic attitude called prejudice.

> To Alan E. Mayers, a Princeton University student,
> October 20, 1950. (Thanks to Mr. Mayers for sending the
> letter to me.) Einstein Archives 83-831

*I am deeply moved by the offer from the State of Israel. . . . All my life I have dealt with objective matters, hence I lack both the natural aptitude and the experience to deal properly with people and to exercise official functions. For these reasons alone I should be unable to fulfill the duties of that high office. . . . My relationship to the Jewish people has become my strongest human bond, ever since I became fully aware of our precarious situation among the nations of the world.

> From a statement to Abba Eban, Israeli ambassador to the
> United States, November 18, 1952, in declining the offer of
> the presidency of Israel. Einstein Archives 28-943

*It also occurred to me that a difficult situation could arise if the government or the parliament were to make decisions that would bring my conscience into conflict; for the fact that one has no actual influence on the course of events does not relieve one of moral responsibility. I am also convinced I would have done a disservice to the cause if I had answered this honorable and tempting call to duty.

To Azriel Carlebach, the editor of *Maariv*, who had pleaded with Einstein to reconsider the offer of the presidency, November 21, 1952. Einstein Archives 41-093

For the young state to achieve real independence, and to conserve it, a group of intellectuals and experts must be produced in the country itself.

From an address recorded on May 11, 1953, for a dinner for Friends of Hebrew University. Quoted in the *New York Times*, May 25, 1953. Einstein Archives 28-987

The Israelis should have chosen English as their language instead of Hebrew. That would have been much better, but they were too fanatical.

Quoted by Fantova, "Conversations with Einstein," January 2, 1954

*For me, the Jewish religion, like all others, is the embodiment of the most childish superstition. And the Jewish people, to whom I gladly belong and with whose way of thinking I share a great affinity, have for me qualities that are no different from all other peoples. In my experience, they are no better than other [ethnic] groups, though they are protected from the worst diseases by a dearth of power. Otherwise I cannot discern anything "chosen" about them.

From a letter to philosopher Eric Gutkind, January 3, 1954. See more in the section "On Religion." The half-page handwritten letter fetched £170,000 ($404,000) at a Bloomsbury auction in London on May 15, 2008, a record for a single Einstein letter and twenty-five times the presale estimate. *New York Times,* May 17, 2008. Einstein Archives 33-337

The German Jews are really terrible, returning to Germany. Even Martin Buber went to Germany and allowed himself to be celebrated with a Goethe Prize [awarded in 1951]. These people are so conceited. I've turned them all down and gave them a kick in the rear.

Fantova, "Conversations with Einstein," February 12, 1954. On February 23, the *New York Times* (p. 5) reported that "Einstein will not join any function of German public life because of crimes of Hitler Germany."

Israel is the only place on Earth where Jews have the possibility to shape public life according to their own traditional ideals.

From an address at a planning conference of American Friends of the Hebrew University, in Princeton, New Jersey, September 19, 1954. Einstein Archives 28-1054

The most important aspect of our [Israel's] policy must be our ever-present, manifest desire to institute complete equality for the Arab citizens living in our midst. . . . The attitude we adopt toward the Arab minority will provide the real test of our moral standards as a people.

To Zvi Lurie, January 4, 1955, written three months before Einstein's death. Einstein Archives 60-388

If I were to be president, sometimes I would have to say to the Israeli people things they would not like to hear.

> To Margot Einstein, on his decision to turn down the presidency of Israel. Quoted in Sayen, *Einstein in America*, 247

*I am *against* nationalism but *in favor* of Zionism. . . . When a man has both arms and he is always saying I have a right arm, then he is a chauvinist. When the right arm is missing, however, then he must do all in his power to make up for the missing limb. Therefore, I am, as a human being, an opponent of nationalism, but as a Jew I support . . . the Jewish-national efforts of the Zionists.

> Quoted by Kurt Blumenfeld, in *Erlebte Judenfrage. Ein Vierteljahrhundert deutscher Zionismus* (Stuttgart: Deutshe Verlags-Anstalt), 127–128

On Life

*Life doesn't make things easy for anyone. But it is lucky when we are able to emerge from our own uncomfortable confines to some extent and focus on objective matters that are beyond the wretchedness of life.

To Adriaan Fokker, July 30, 1919. *CPAE*, Vol. 9, Doc. 78

The finest things in life include having a clear grasp of correlations. One can deny this only in a very dismal, nihilistic mood.

To Hedwig Born, August 31, 1919. *CPAE*, Vol. 9, Doc. 97

If there is no price to be paid, it is also not of value.

Aphorism, June 27, 1927. Einstein Archives 35-582

*Life in the service of an idea can be good if this idea is life-giving and emancipates the individual from the fetters of the self without propelling him into a different bondage. Science and art *can* work this way, but they can also lead to enslavement or unhealthy pampering and overrefinement. But I would not dispute the notion that these efforts lead to an inability to cope with life. After all, even water is a poison if you drown in it.

To son Eduard, December 23, 1927. Translated in Neffe, *Einstein*, 194. Einstein Archives 75-748

*If A is success in life, then $A = x + y + z$. Work is x, play is y, and z is keeping your mouth shut.

> Said to Samuel J. Woolf in Berlin, Summer 1929. Published in the *New York Times*, August 18, 1929. Einstein often spoke about keeping one's mouth shut. See, for example, his remarks about listening to music in the section "On Music."

*Sometimes one pays most for things one gets for nothing.

> From an interview with G. S. Viereck, "What Life Means to Einstein," *Saturday Evening Post*, October 26, 1929; reprinted in Viereck, *Glimpses of the Great*, 434

Life is a great tapestry. The individual is only an insignificant thread in an immense and miraculous pattern.

> Ibid., 444

Strange is our situation here on Earth. Each of us comes for a short visit, not knowing why, yet sometimes seeming to divine a purpose. . . .There is one thing we do know: that man is here for the sake of other men.

> Opening words of "What I Believe," *Forum and Century* 84 (1930), 193–194. See Rowe and Schulmann, *Einstein on Politics*, 226–230, for the whole essay and background information. Different versions of the essay have appeared in various publications, sometimes under the title "My Credo."

Only a life lived for others is a life worthwhile.

> In answer to a question asked by the editors of *Youth*, a journal of Young Israel of Williamsburg, N.Y. Quoted in the *New York Times*, June 20, 1932. Einstein Archives 29-041

The life of the individual has meaning only insofar as it aids in making the life of every living thing nobler and more beautiful. Life is sacred, that is to say, it is the supreme value, to which all other values are subordinate.

> From "Is There a Jewish Point of View?"August 3, 1932. Published in *Mein Weltbild* (1934), 89–90; reprinted in *Ideas and Opinions*, 185–187

Every reminiscence is colored by today's being what it is, and therefore by a deceptive point of view.

> Written in 1946 for "Autobiographical Notes," 3

*Life is short, and the boulder against which one pushes with all one's might moves from its spot only with long intermissions.

> To Hans Muehsam, January 22, 1947. Einstein Archives 38-361

Personal existence makes sense through the conviction of the objective value of one's own strife and action. But if this conviction is not softened by humor, one becomes insufferable.

> To Johanna Fantova, October 9, 1948. One of three aphorisms sent to her. Einstein Archives 87-034

A life directed chiefly toward the fulfillment of personal desires will sooner or later always lead to bitter disappointment.

To T. Lee, January 16, 1954. Einstein Archives 60-235

If you want to live a happy life, tie it to a goal, not to people or objects.

Quoted by Ernst Straus in French, *Einstein: A Centenary Volume*, 32

On Music

Bach, Mozart, and some old Italian and English composers were Einstein's favorites; also Schubert, because of the composer's ability to express emotion. He was considerably less fond of Beethoven, regarding his music as too dramatic and personal. Handel, he felt, was technically good but displayed shallowness. Schumann's shorter works were appealing because they were original and rich in feeling. Mendelssohn demonstrated considerable talent but lacked depth. Einstein liked some lieder and chamber music by Brahms. He found Wagner's musical personality indescribably offensive "so that for the most part I can listen to him only with disgust." He considered Richard Strauss gifted but without inner truth and concerned too much with outside effect. (From a response to a questionnaire, May 1939. Einstein Archives 34-322)

Einstein began to play the violin at age six; by the mid-1940s he had given it up and played around only on the piano. According to Barbara Wolff of the Einstein Archives, Einstein had at least ten different violins between 1920 and 1950. He allegedly called at least one of his violins "Lina" and willed his last one to his grandson, Bernhard, who gave it to his son Paul. See Frank, *Einstein: His Life and Times*, 14; Grüning, *Ein Haus für Albert Einstein*, 251.

Stick to the Mozart sonatas. Your papa, too, learned to know music well through them.

To son Hans Albert, January 8, 1917. Einstein had "fallen in love with the Mozart sonatas" at age thirteen. *CPAE*, Vol. 8, Doc. 287, n. 2

The differences between Japanese music and our own are truly fundamental. Whereas in our European music, chords and architectural arrangement are essential and are a given, they are absent in Japanese music. Both, however, have the same thirteen notes that make up an octave. To me, Japanese music is a painting of emotions that has a surprising and immediate effect. . . . I have the impression that it is all about giving a stylized presentation of the emotions found in the human voice, as well as the sounds of nature that stir the human soul, such as birdsong and the rumble of the ocean. This sensation is given force because percussion instruments, which are not limited by pitch and are especially well suited for rhythmic characterization, play a large role. . . . To my mind, the greatest obstacle to the acceptance of Japanese music as a great art form is its lack of formal arrangement and architectural structure.

From "My Impressions of Japan," *Kaizo* 5, no. 1 (January 1923), 339. Einstein Archives 36-477.1

Unfortunately I don't feel I am in a position, on the strength of either my sexual or my musical abilities, to accept your kind invitation.

To Kurt Singer, August 16, 1926, in declining an invitation to participate in a musical event at the First International Congress for Sexual Research. Einstein was playing with words, as he had been invited to play one of the violin parts of the Brahms String Sextet, no. 1 B-dur, op. 18. Einstein Archives 44-905

Music does not influence research work, but both are nourished by the same sort of longing, and they complement each other in the satisfaction they offer.

> To Paul Plaut, October 23, 1928. Quoted in Dukas and Hoffmann, *Albert Einstein, the Human Side*, 78. Einstein Archives 28-065

As to Schubert, I have only this to say: play the music, love—and keep your mouth shut!

> Reply to yet another question on composers, November 10, 1928. Quoted in Dukas and Hoffmann, *Albert Einstein, the Human Side*, 75

This is what I have to say about Bach's life and work: listen, play, love, revere—and keep your mouth shut.

> Reply to a questionnaire on Bach for the magazine *Reclams Universum*, 1928. Quoted in *Einstein: A Portrait*, 74, and Dukas and Hoffmann, *Albert Einstein, the Human Side*, 75. Einstein Archives 28-058.1

If I were not a physicist, I would probably be a musician. I often think in music. I live my daydreams in music. I see my life in terms of music. . . . I get most joy in life out of my violin.

> From an interview with G. S. Viereck, "What Life Means to Einstein," *Saturday Evening Post*, October 26, 1929; reprinted in Viereck, *Glimpses of the Great*, 436

In Europe, music has come too far away from popular art and popular feeling and has become something like a secret art with conventions and traditions of its own.

From a conversation with Indian mystic, poet, and musician Rabindranath Tagore on August 19, 1930, in Berlin, discussing the possibility of self-expression in Eastern and Western music. In *Asia* 31 (March 1931), 140–142

It requires a very high standard of art to realize fully a great idea in original music so that one can make variations upon it. [In the West], the variations are often prescribed.

Ibid.

The difficulty is that really good music, whether of the East or of the West, cannot be analyzed.

Ibid.

*The unemployment situation of musicians in general is appalling, not so much in destitute nations as in those with a meager musical culture where children no longer learn to play instruments.

To A. Woehr, April 5, 1933. Einstein Archives 52-305

*Don't read any newspapers, find a few like-minded people and read the wonderful writers of the past, Kant, Goethe, Lessing, and the classics of other countries, and take joy in the wonderful natural environ-

ment you can find in Munich. . . . Seek the compan-
ionship of a few animals.

Ibid. Woehr was an unemployed musician and asked Einstein for advice on how to lead his life.

Mozart's music is so pure and beautiful that I see it
as a reflection of the inner beauty of the universe.

As recalled by Peter Bucky, *The Private Albert Einstein* (1933)

I took violin lessons from age 6 to 14, but had no
luck with my teachers, for whom music did not
transcend mechanical practicing. I really began to
learn only when I was about 13 years old, mainly
after I had fallen in love with Mozart's sonatas.

In draft of letter to Philipp Frank, 1940. Einstein Archives
71-191

I feel uncomfortable listening to Beethoven. I think
he is too personal, almost naked. Give me Bach,
rather, and then more Bach.

From an interview with Lili Foldes, *The Etude*, January 1947

My knowledge of modern music is very restricted.
But in one respect I feel certain: true art is characterized by an unstoppable urge in the creative artist.

From a tribute to Ernst Bloch, November 15, 1950. Quoted
in Dukas and Hoffmann, *Albert Einstein, the Human Side*, 77.
Einstein Archives 34-332

I am done fiddling. With the passage of years, it has become more and more unbearable for me to listen to my own playing.

> To Queen Elisabeth of Belgium, January 6, 1951. Quoted in Nathan and Norden, *Einstein on Peace*, 554. Einstein Archives 32-400

The piano is much more suited to improvisation [than the violin], also for playing alone, and I play the piano every day. Besides, it would be much too physically strenuous for me to play the violin now.

> Quoted by Fantova, "Conversations with Einstein," March 24, 1954

Today I heard Mozart's *Jupiter* Symphony on the radio. It is the best piece that Mozart wrote. Of Mozart's operas, *Figaro* and *The Abduction* [*from the Seraglio*] are outstanding. I don't like *The Magic Flute* so much. Of the modern ones, only Mussorgsky's *Boris Godunov* is good.

> Ibid., March 10, 1955

Mozart wrote such nonsense here!

> While struggling to play a piece by Mozart. Quoted by Margot Einstein in an interview with Jamie Sayen for *Einstein in America*, 139

First I improvise, and if that doesn't help, I seek solace in Mozart. But when I am improvising and it appears that something may come of it, I require the clear constructions of Bach in order to follow through.

> Explaining how he relaxes after work playing his violin in his Berlin kitchen, a room with superior acoustics. Recalled by Konrad Wachsmann in Grüning, *Ein Haus für Albert Einstein*, 251. Quoted in Ehlers, *Liebes Hertz!* 132

On Pacifism, Disarmament, and World Government

Einstein was a pacifist from his youth until 1933, when Hitler forced his hand on the issue. From 1933 to 1945, he saw some need for military action under certain circumstances; in particular, he felt that military strength among "the nations that have stayed normal" was vital against the German aggressor (Fölsing, *Albert Einstein*, 676). In general, however, he believed that a "supranational" world government for the control of weapons was necessary to preserve civilization and individual freedoms. From 1945 until his death in 1955, he spoke out in favor of world government as a moral imperative.

*An affinity for the political structure of Germany would be unnatural for me as a pacifist.

To Fritz Haber, March 9, 1921. *CPAE*, Vol. 12, Doc. 87

He who cherishes the values of culture cannot fail to be a pacifist.

From Einstein's contribution to *Die Friedensbewegung*, ed. Kurt Lenz and Walter Fabian (1922). Quoted in Nathan and Norden, *Einstein on Peace*, 55

Most of those representing the field of history have certainly done little to foster the cause of pacifism. Many of its representatives . . . have publicly made astounding and strongly chauvinistic and military pronouncements. . . . The situation is quite different in the natural sciences.

From *Die Friedensbewegung*, ibid.

Because of the universal character of their subject matter and their need for internationally organized cooperation, [scientists] are inclined toward international understanding and therefore favor pacifist goals.

Ibid.

Technology resulting from the sciences has internationally chained together economies and this has caused all wars to become a matter of international importance. When this situation has entered the consciousness of mankind, after sufficient turmoil, then men will also find the energy and goodwill to create organizations that have the power to end wars.

Ibid.

I wish (1) that next year will bring the broadest possible international agreements on disarmament on land and at sea; (2) that a solution will be found for the international war debts that allows the European states to pay their obligations without having to pawn their property abroad; (3) that an honest arrangement can be reached with the Soviet Union that frees this land from external pressures while allowing its internal development to proceed unhindered.

From a statement for the December 31, 1928, *Chicago Daily News*, after being asked by reporter Edgar Mowrer what his wishes were for the New Year. (Courtesy of Uriel Gorney and Mishael Zedek.) Einstein Archives 47-670

No person has the right to call himself a Christian or Jew so long as he is prepared to engage in systematic murder at the command of an authority, or allow himself to be used in any way in the service of war or the preparation for it.

From a statement for "Livre d'or de la paix," 1928. Quoted in Nathan and Norden, *Einstein on Peace*. Einstein Archives 28-054

I would absolutely refuse any direct or indirect war service and would try to persuade my friends to do the same, regardless of the reasons for the cause of a war.

Written February 23, 1929, for publication in *Die Wahrheit* (Prague, 1929). Also in Nathan and Norden, *Einstein on Peace*. Einstein Archives 48-684

My pacifism is an instinctive feeling, a feeling that possesses me because the murder of people is disgusting. My attitude is not derived from any intellectual theory but is based on my deepest antipathy to every kind of cruelty and hatred.

To Paul Hutchinson, editor of *Christian Century*, July 1929. An interview was published in *Christian Century*, August 28, 1929. Quoted in Nathan and Norden, *Einstein on Peace*, 98

I have made no secret, either privately or publicly, of any sense of outrage over officially enforced military and war service. I regard it as a duty of conscience to fight against such barbarous enslavement of the individual with every means available.

> From a statement to the Danish newspaper *Politiken*, August 5, 1930. Reprinted in Nathan and Norden, *Einstein on Peace*, 129. Einstein Archives 48-036

*When those who are bound together by pacifist ideals hold a meeting they are always consorting with their own kind only. They are like sheep huddled together while the wolves wait outside. . . . The sheep's voice does not get beyond this circle and therefore is ineffective.

> From the "Two Percent" speech for the New History Society, an offshoot of the pacifist Baha'i religion, in New York, December 14, 1930. From notes taken by Rosika Schwimmer. See Rowe and Schulmann, *Einstein on Politics*, 240. Einstein Archives 48-479

*Under our present system of military duty everyone is compelled to commit a crime—the crime of killing people for his country. The aim of all pacifists must be to convince others of the immorality of war and rid the world of the shameful slavery of military service.

> Ibid.

*Even if only two percent of those supposed to perform military service should declare themselves war resisters and assert, "We are not going to fight.

We need other methods of settling international disputes," the governments would be powerless—they could not put such masses into jail.

Ibid., 241

*The man who enjoys marching in line and file to the strains of music falls below my contempt. This heroism on command, this senseless violence, this accursed bombast of patriotism—how intensely I despise them! War is low and despicable, and I had rather be smitten to shreds than participate in such doings.

From "What I Believe," *Forum and Century* 84 (1930), 193–194. Variously translated elsewhere, including in earlier editions of this book. Reprinted in Rowe and Schulmann, *Einstein on Politics*, 229

I believe serious progress [in the abolition of war] can be achieved only when men become organized on an international scale and refuse, as a body, to enter military or war service.

From a statement in *Jugendtribüne*, April 17, 1931. Einstein Archives 47-165

*Few of us still cling to the notion that acts of violence in the shape of wars are either advantageous or worthy of humanity as a method of solving international problems. But we are not consistent enough to make vigorous efforts on behalf of the measures which might prevent war, that savage and unworthy relic of the age of barbarism.

From "America and the Disarmament Conference of 1932,"
June 1931. May have been delivered at Whittier College in
California on January 18, 1932, according to Nathan and
Norden, *Einstein on Peace*, 658. Reprinted in Rowe and
Schulmann, *Einstein on Politics*, 248. Einstein Archives
28-152

There are two ways of resisting war: the legal way and the revolutionary way. The legal way involves the offer of alternative service not as a privilege for a few but as a right for all. The revolutionary view involves an uncompromising resistance, with a view to breaking the power of militarism in time of peace or the resources of the state in time of war.

From a statement recorded by Fenner Brockway for *The
New World*, July 1931, after meeting with Einstein in May
1931. Einstein Archives 47-742

I appeal to all men and women, whether they be eminent or humble, to declare that they will refuse to give any further assistance to war or the preparation of war.

From a statement to the War Resisters International, Lyon,
France, August 1931. Quoted in Frank, *Einstein: His Life and
Times*, 158; also quoted in the *New York Times*, August 2,
1931

*Anybody who really wants to abolish war must resolutely declare himself in favor of his own country's resigning a portion of its sovereignty in favor of international institutions: he must be ready to make his own country amenable, in case of a dispute, to the award of an international court. He must . . . support disarmament all round.

Ibid. Einstein uses almost the same words in a letter to Sigmund Freud of July 30, 1932 (Einstein Archives 32-543, 545, 546). He returns to this theme of world government after World War II.

*The State exists for man, not man for the State. . . . I believe that the most important mission of the State is to protect the individual and make it possible for him to develop into a creative personality. The State should be our servant; we should not be slaves of the State. The State violates this precept when it forces us to perform military service.

From "The Road to Peace," published in the *New York Times Magazine*, November 22, 1931. Reprinted in Rowe and Schulmann, *Einstein on Politics*, 253. (In previous editions of *The Quotable Einstein*, I had given a slightly different version and a wrong reference used in *Ideas and Opinions*.) Einstein Archives 28-175

*To refuse on moral grounds to perform military service may expose one to severe persecution; is this persecution any less shameful for society than the persecution to which the religious martyrs were subjected in earlier centuries?

Ibid., 255

I am not only a pacifist, but a militant pacifist. I am willing to fight for peace. . . . Is it not better for a man to die for a cause in which he believes, such as peace, than to suffer for a cause in which he does not believe, such as war?

> From an interview with G. S. Viereck in 1931, published in a booklet with other texts, *The Fight against War*, ed. Alfred Lief (New York: John Day, 1933). As quoted in Nathan and Norden, *Einstein on Peace*, 125–126

Peace cannot be kept by force. It can only be achieved by understanding. You cannot subjugate a nation forcibly unless you wipe out every man, woman, and child. Unless you wish to use such drastic measures, you must find a way of settling your disputes without resort to arms.

> From "Notes on Pacifism," in *Cosmic Religion* (1931), 67. Probably a paraphrase. Original source unknown.

I am the same ardent pacifist I was before. But I believe that the tool of refusing military service can be advocated again in Europe only when the military threat from aggressive dictatorships toward democratic countries has ceased to exist.

> To Rabbi Philip Bernstein, April 5, 1934. In Nathan and Norden, *Einstein on Peace*, 250. Einstein Archives 49-276

Only if we succeed in abolishing compulsory military service altogether will it be possible to educate the youth in the spirit of reconciliation, joy in life, and love toward all living creatures.

From "Three Letters to Friends of Peace," published in *Mein Weltbild* (1934); reprinted in *Ideas and Opinions*, 109

*To arm is to give one's voice and make one's preparations, not for peace but for war. Therefore people will not disarm step by step; they will disarm at one blow or not at all.

From "The Question of Disarmament," published in ibid.; translated slightly differently in *Ideas and Opinions*, 102–103. Taken from Rowe and Schulmann, *Einstein on Politics*, 22. Einstein Archives 28-180

*I stand firmly by the principle that a real solution of the problem of pacifism can be achieved only by the organization of a super-national court of arbitration, which, differing from the present League of Nations in Geneva, would have at its disposal the means of enforcing its decisions. In short, an international court of justice with a permanent military establishment—or better, police force.

From "A Re-examination of Pacifism," *Polity* 3, no. 1 (January 1935), 4–5. Reprinted in Rowe and Schulmann, *Einstein on Politics*, 284–286. Einstein Archives 28-296

Pacifism defeats itself under certain conditions, as it would in Germany today. . . . We must work with the people to create a public sentiment that will outlaw

war: (1) create the idea of supersovereignty; . . .
(2) face the economic causes of war.

> From an interview with Robert M. Bartlett, *Survey Graphic*
> 24 (August 1935), 384, 413. Reprinted in Nathan and Nor-
> den, *Einstein on Peace*, 260

I am convinced that an international political organization is not only possible but is unconditionally
necessary if the situation on our planet should eventually become unbearable.

> From a draft manuscript, ca. 1940. See Kaller's autographs
> catalog, "Jewish Visionaries," 35

In the twenties, when no dictatorships existed, I advocated that refusing to go to war would make war
improper. But as soon as coercive conditions appeared in certain nations, I felt that it would weaken
the less aggressive nations vis-à-vis the more aggressive ones.

> From an interview, *New York Times*, December 30, 1941.
> Einstein Archives 29-096

There is no other salvation for civilization and even
for the human race than in the creation of a world
government, with the security of nations founded
upon law. As long as there are sovereign states with
their separate armaments and armament secrets,
new world wars cannot be avoided.

> From a press interview, *New York Times*, September 15, 1945

*At the present high level of industrialization and economic interdependence, it is unthinkable that we can achieve peace without a genuine supranational organization to govern international relations. If war is to be avoided, anything less than such an over-all solution strikes me as illusory.

To J. Robert Oppenheimer, September 29, 1945. Einstein Archives 57-294

*The secret of the bomb should be committed to a world government, and the United States should immediately announce its readiness to do so. Such a world government should be established by the United States, the Soviet Union and Great Britain, the only three powers which possess great military strength. The three of them should commit to this world government all of their military resources. The fact that there are only three nations with great military power should make it easier, rather than harder, to establish a world government.

To Raymond Swing, *Atlantic Monthly* 176, no. 5 (November 1945), 43–45

Everything that is done in international affairs must be done from the following viewpoint: Will it help or hinder the establishment of world government?

From the text of a broadcast interview with P. A. Schilpp and F. Parmelee, May 29, 1946. See also Nathan and Norden, *Einstein on Peace*, 382. Einstein Archives 29-105

A world government must be created which is able to solve conflicts between nations by judicial decision. . . . This government must be based on a clear-cut constitution that is approved by the governments and the nations, and which has the sole disposition of offensive weapons.

From a broadcast for a rally of Students for Federal World Government, Chicago, May 29, 1946. See *New York Times*, May 30, 1946. Quoted in Pais, *Einstein Lived Here*, 232. Einstein Archives 28-694

I advocate world government because I am convinced that there is no other possible way of eliminating the most terrible danger in which man has ever found himself. The objective of avoiding total destruction must have priority over any other objective.

From "A Reply to the Soviet Scientists," *Bulletin of Atomic Scientists*, February 1948. Also in *Ideas and Opinions*, 140–146. Einstein Archives 28-795

There is only one path to peace and security: the path of a supranational organization. One-sided armament on a national basis only heightens the general uncertainty and confusion without being an effective protection.

From an address at Carnegie Hall in New York on receiving the One World Award, April 27, 1948. Published in *Out of My Later Years*; reprinted in *Ideas and Opinions*, 147

If the idea of world government is not realistic, then there is only one realistic view of our future: wholesale destruction of man by man.

From a comment about the film *Where Will You Hide?* May 1948. Einstein Archives 28-817

Mankind can be saved only if a supranational system, based on law, is created to eliminate the methods of brute force.

From a statement in *Impact* 1 (1950), 104. Einstein Archives 28-882

It is my belief that the problem of bringing peace to the world on a supranational basis will be solved only by employing Gandhi's method on a larger scale.

To Gerhard Nellhaus, March 20, 1951. Einstein Archives 60-684

The conscientious objector is a revolutionary. In deciding to disobey the law he sacrifices his personal interests to the most important cause of working for the betterment of society.

Ibid.

I can identify my views almost completely with those of Gandhi. But I would (individually and collectively) resist with violence any attempt to kill or to take away from my people or me the basic means of subsistence.

To A. Morrisett, March 21, 1952. Einstein Archives 60-595

The goal of pacifism is possible only through a supranational organization. To stand unconditionally for this cause is . . . the criterion of true pacifism.

Ibid.

I believe that the killing of human beings in a war is no better than common murder.

To the editor of the Japanese magazine *Kaizo*, September 20, 1952. See Rowe and Schulmann, *Einstein on Politics*, 488. Einstein Archives 60-039

My participation in the production of the atomic bomb consisted of one single act: I signed a letter to President Roosevelt in which I emphasized the necessity of conducting large-scale experimentation with regard to the feasibility of producing an atom bomb. . . . I felt impelled to take the step because it seemed probable that the Germans might be working on the same problem with every prospect of success. I had no alternative to act as I did, *although I have always been a convinced pacifist*.

Ibid.

The more a country makes military weapons, the more insecure it becomes: if you have weapons, you become a target for attack.

> From an interview with A. Aram, January 3, 1953. Einstein Archives 59-109

In my letter to *Kaizo*, I did not say that I was an *absolute* pacifist, but rather that I had always been a *convinced* pacifist. While I am a convinced pacifist, there are circumstances in which I believe the use of force is appropriate—namely, in the face of an enemy unconditionally bent on destroying me and my people.

> To Japanese pacifist Seiei Shinohara, February 22, 1953. Shinohara, who felt Gandhi would not have written the letter to Roosevelt if he were in Einstein's position, had not accepted Einstein's remarks in *Kaizo* as valid, prompting the above reply. See Rowe and Schulmann, *Einstein on Politics*, 490. Einstein Archives 61-295

I am a *dedicated* but not an *absolute* pacifist; this means that I am opposed to the use of force under any circumstances except when confronted by an enemy who pursues the destruction of life as an *end in itself*.

> To Seiei Shinohara, June 23, 1953. See Rowe and Schulmann, *Einstein on Politics*, 491. Einstein Archives 61-297

In all cases where a reasonable solution of difficulties is possible, I favor honest cooperation and, if this is not possible under prevailing circumstances, Gandhi's method of peaceful resistance to evil.

To John Moore, November 9, 1953. Quoted in Nathan and Norden, *Einstein on Peace*, 596. Einstein Archives 60-584

I have always been a pacifist, i.e., I have declined to recognize brute force as a means for the solution of international conflicts. Nevertheless, it is, in my opinion, not reasonable to cling to that principle unconditionally. An exception has necessarily to be made if a hostile power threatens wholesale destruction of one's own group.

To H. Herbert Fox, May 18, 1954. Einstein Archives 59-727

On Peace, War, the Bomb, and the Military

*The struggle raging today will likely produce no victor. . . . Therefore, it seems not only ethically fitting, but rather bitterly necessary that intellectuals of all nations marshal their influence such that the terms of peace shall not become the cause of future wars.

From "Manifesto to the Europeans," prepared in mid-October 1914 by Georg Nicolai, Wilhelm Foerster, and Einstein to counter an earlier manifesto upholding German justification for actions in the early phases of World War I. *CPAE*, Vol. 6, Doc. 8

Even the scholars in various lands have been acting as if their brains had been amputated.

To Romain Rolland, March 22, 1915, on the outbreak of World War I. Rolland was the most prominent pacifist of his time. *CPAE*, Vol. 8, Doc. 65

The psychological roots of war are, in my opinion, biologically rooted in the aggressive nature of the male creature. . . . Some animals—the bull and the rooster—surpass us in this regard.

From "My Opinion on the War," for the Goethebund of Berlin, October–November 1915. *CPAE*, Vol. 6, Doc. 20. Einstein repeated this opinion in an interview with anthropologist Ashley Montagu thirty-one years later, in June 1946, claiming that a child's naughtiness and a parent's spanking—"domestic violence"—were innately reactive, instinctive acts and a microcosmic example of international violence and aggression. Einstein was essentially agreeing with Sigmund Freud's conclusions, but Montagu disagreed and persuaded Einstein that the doctrine of man's innate depravity was unsound. See Montagu's article, "Conversations with Einstein," in *Science Digest*, July 1985. Einstein Archives 29-002

*The internationalism that existed before the war, before 1914, the internationalism of culture, the cosmopolitanism of commerce and industry, the broad tolerance of ideas—this internationalism was essentially right. There will be no peace on earth, the wounds inflicted by the war will not heal, until this internationalism is restored.

From an interview in the *New York Evening Post*, March 26, 1921. See Rowe and Schulmann, *Einstein on Politics*, 89

*In spite of so many hopes and illusions, war will always be possible. The world does not seem to fear even the most extreme and catastrophic inhumanity and murderousness of war.

From an interview with Aldo Sorani in *Il Messagero*, October 26, 1921. *CPAE*, Vol. 12, Appendix G

*I think that the greatest contribution to peace could be made by the press, [which] too often helps the war or helps to stir up political unrest. If the press in all countries were united in a plan of peace, a decisive step would be taken toward the achievement of our ideal of harmony, brotherhood, and universal sharing of goods in the world.

Ibid.

*The people in every country will insist that their own nation is the victim of aggression and will do so in perfectly good faith. . . . You cannot educate a nation for war and, at the same time, make its people believe that war is a shameful crime.

To Jacques Hadamard, September 24, 1929. Einstein Archives 12-025

*I admit that the country which decides not to defend itself assumes a great risk. However, this risk is accepted by society as a whole, and in the interest of human progress. Real progress has never been possible without sacrifices. . . . As long as nations systematically continue to prepare for war, fear, distrust and selfish ambitions will lead to war again.

Ibid.

*To wage war means both to kill the innocent and to allow oneself to be innocently killed. . . . How can any decent and self-respecting person participate in such a tragic affair? Would you perjure yourself if your government asked you to do so? Certainly not. How much worse, then, to slaughter innocent men?

Ibid.

As long as armies exist, any serious conflict will lead to war. A pacifism that does not actively fight against the armament of nations is and must remain impotent.

From "Active Pacifism," August 8, 1931, for a peace demonstration at Dixmuide, a Belgian battleground in World War I. Published in *Mein Weltbild* (1934), 55; reprinted in *Ideas and Opinions*, 111

May the conscience and the common sense of the people be awakened, so that we may reach a new stage in the life of nations, where people will look back on war as an incomprehensible aberration of their forefathers!

Ibid.

War is not a parlor game in which the players obediently stick to the rules. Where life and death are at stake, rules and obligations go by the board. Only the absolute repudiation of all war can be of any use here.

From an address to university students in California, February 27, 1932. Source misquoted in *Ideas and Opinions*, 93. Published in the *New York Times*, February 28, 1932. Einstein Archives 28-187

*Moral disarmament, like the problem of peace as a whole, is made difficult because men in power never want to surrender any part of their country's sovereignty, which is exactly what they must do if war is to be abolished.

From a press conference at the Geneva Disarmament Conference, May 23, 1932. See Rowe and Schulmann, *Einstein on Politics*, 257, who took this version from the German edition of Nathan and Norden, *Einstein on Peace*, or 168–169 of the English edition. Nathan and Norden note that Einstein revised another version of the transcript of the conference, prepared by War Resisters International, because he felt he had been misunderstood. They state that "Einstein's own version" (Einstein Archives 72-559) was used as the basis for their English translation.

*I am absolutely convinced that we should use every possible means to strengthen the war resistance movement. Its moral significance cannot be overestimated. . . . [It] inspires individual courage, challenges the conscience of men, and undermines the authority of the military system.

Ibid.

*This is not a comedy. This is a tragedy . . . despite the cap and bells and buffoonery. No one has any right to treat this tragedy lightly or to laugh when one should cry. We should be standing on rooftops, all of us, and denouncing this conference as a travesty!

From an interview by Konrad Bercovici before the press conference referred to above. Published in *Pictorial Review* the following year, February 1933; quoted in Clark, *Einstein*, 372. Clark states that "no one would claim that this is necessarily an accurate . . . verbatim account of what Einstein said. But from all available evidence it would seem to reflect his excited . . . attitude."

We must . . . dedicate our lives to drying up the source of war: ammunition factories.

> Ibid., 373

This is the problem: Is there any way of delivering mankind from the menace of war? It is common knowledge that with the advance of modern science, this issue has come to mean a matter of life and death for civilization as we know it; nevertheless, for all the zeal displayed, every attempt at its solution has ended in a lamentable breakdown.

> To Sigmund Freud, July 30, 1932. Published, along with Freud's reply, as *Why War?* by the League of Nations. Also quoted in Nathan and Norden, *Einstein on Peace*, 188. Einstein Archives 32-543

Anyone who really wants to abolish war must resolutely declare himself in favor of his own country's resigning a portion of its sovereignty in favor of international institutions.

> From "America and the Disarmament Conference of 1932." Published in *Mein Weltbild*, 63; reprinted in *Ideas and Opinions*, 101

Compulsory military service, as a hotbed of unhealthy nationalism, must be abolished; most important, conscientious objectors must be protected on an international basis.

From "The Disarmament Conference of 1932," in *Ideas and Opinions*, 98. A different version appears in *The Nation*, Vol. 133, 300

Compulsory military service seems to me the most disgraceful symptom of the deficiency in personal dignity from which civilized mankind is suffering today.

From "Society and Personality," 1932. Published in *Mein Weltbild* (1934); reprinted in *Ideas and Opinions*, 15

The powerful industrial groups concerned with the manufacture of arms are doing their best in all countries to prevent the peaceful settlement of international disputes, and rulers can achieve this great end only if they are sure of the vigorous support of the majority of their people. In these days of democratic government, the fate of nations hangs on the people themselves; each individual must always bear that in mind.

From "Peace," 1932. Published in *Mein Weltbild* (1934); reprinted in *Ideas and Opinions*, 106

*Without doubt the present economic difficulties will bring forth some legislation to the effect that an adjustment between supply and demand of labor as well as between production and consumption will always be brought about through government control. But these problems too must be solved by free men.

Ibid.

It is unworthy of a great nation to stand idly by while small countries of great culture are being destroyed with a cynical contempt for justice.

> From a message to a peace meeting at Madison Square Garden, April 5, 1938. Quoted in Nathan and Norden, *Einstein on Peace*, 279. Einstein Archives 28-424

*Some recent work by E. Fermi and L. Szilard, which has been communicated to me in manuscript, leads me to expect that the element uranium may be turned into a new and important source of energy in the immediate future. Certain aspects of the situation seem to call for watchfulness and, if necessary, quick action on the part of the Administration.

> The first paragraph of Einstein's famous letter to President Roosevelt concerning his fear that Germany might build a bomb, August 2, 1939. Widely available on the Internet. Einstein was not privy to any further developments and, lacking security clearance, did not participate in the Manhattan Project that built the American bomb. He did consult on other, more minor, military matters. Einstein Archives 33-088

Organized power can be opposed only by organized power. Much as I regret this, there is no other way.

> To R. Fowlkes, a pacifist student, July 14, 1941. Einstein Archives 55-100

*The terms of secrecy under which Dr. Szilard is working at present do not permit him to give me information about his work; however, I understand that he now is greatly concerned about the lack of adequate contact between scientists who are doing this work and those members of your Cabinet who are responsible for formulating policy. In the circumstances, I consider it my duty to give Dr. Szilard this introduction and I wish to express the hope that you will be able to give his presentation of the case your personal attention.

To President Roosevelt, March 25, 1945, after Einstein became concerned about the control of atomic energy and the political policy implications. Roosevelt died on April 12 and never saw the letter. Einstein Archives 33-109

I have done no work on [the atomic bomb], no work at all. I am interested in the bomb the same as any other person, perhaps a little bit more interested.

From an interview with Richard J. Lewis, *New York Times*, August 12, 1945

*Dear Albert! My scientific work has no more than a very indirect connection to the atomic bomb.

From a letter of assurance to his son Hans Albert, September 2, 1945. Translated in Neffe, *Einstein*, 388. Einstein Archives 75-790

As long as nations demand unrestricted sovereignty we shall undoubtedly be faced with still bigger wars, fought with bigger and technologically more advanced weapons.

To Robert Hutchins, September 10, 1945. Quoted in Nathan and Norden, *Einstein on Peace*, 337. Einstein Archives 56-894

The release of atomic energy has not created a new problem. It has merely made more urgent the necessary solving of an existing one. One could say it has affected us quantitatively, not qualitatively.

From "Einstein on the Atomic Bomb," part 1, as told to Raymond Swing, *Atlantic Monthly* 176, no. 5 (November 1945), 43–45. See Rowe and Schulmann, *Einstein on Politics*, 373–378

I do not believe that civilization will be wiped out in a war fought with the atomic bomb. Perhaps two-thirds of the people on Earth would be killed, but enough men capable of thinking, and enough books, would be left to start out again, and civilization would be restored.

Ibid.

The secret of the bomb should be committed to a world government. . . . Do I fear the tyranny of a world government? Of course I do. But I fear still more the coming of another war or wars. Any government is certain to be evil to some extent. But a world government is preferable to the far greater evil of wars.

Ibid.

I do not consider myself the father of the release of atomic energy. My part in it was quite indirect. I did not, in fact, foresee that it would be released in my lifetime. I believed only that it was theoretically possible. It became practical only through the accidental discovery of a chain reaction, and this was not something I could have predicted.

Ibid.

The war is won, but the peace is not. The great powers, united in fighting, are now divided over the peace settlements.

From an address at the fifth Nobel anniversary dinner in New York, December 10, 1945. See Rowe and Schulmann, *Einstein on Politics*, 382. Einstein Archives 28-722

*Many persons have inquired concerning a recent message of mine that "a new type of thinking is essential if mankind is to survive and move to higher levels." . . . Past thinking and methods did not prevent world wars. Future thinking *must* prevent wars.

Ibid., 383. This is one of the most queried Einstein quotations.

*In previous ages a nation's life and culture could be protected to some extent by the growth of armies in national competition. Today we must abandon competition and secure cooperation.

Ibid.

Rifle bullets kill men, but atomic bombs kill cities. A tank is a defense against a bullet but there is no defense in science against the weapon which can destroy civilization. . . . Our defense is in law and order.

Ibid., 384

The unleashed power of the atom bomb has changed everything except our modes of thinking, and thus we drift toward unparalleled catastrophes.

From a letter for the Emergency Committee of Atomic Scientists, May 23, 1946. Quoted in Nathan and Norden, *Einstein on Peace*, 376. Einstein Archives 88-539

Science has brought forth this danger, but the real problem is in the minds and hearts of men. We will not change the hearts of other men by mechanisms, but by changing *our* hearts and speaking bravely. . . . When we are clear in heart and mind—only then shall we find courage to surmount the fear which haunts the world.

On nuclear weapons. From interview with Michael Amrine, "The Real Problem Is in the Hearts of Men," *New York Times Magazine*, June 23, 1946. See Rowe and Schulmann, *Einstein on Politics*, 387–388

Non-cooperation in military matters should be an essential moral principle for all true scientists . . . who are engaged in basic research.

In answer to a question posed by the Overseas News
Agency, January 20, 1947. Quoted in Nathan and Norden,
Einstein on Peace, 401. Einstein Archives 28-733

*Through the release of atomic energy, our genera-
tion has brought into the world the most revolutionary force since man's discovery of fire.

In a letter of support for the Emergency Committee of
Atomic Scientists, January 22, 1947. Many such letters were
sent under various dates. Einstein Archives 40-010

*We scientists recognize our inescapable responsi-
bility to carry to our fellow citizens an understanding of the simple facts of atomic energy and its implications for society. In this lies our only security
and our only hope.

Ibid.

Had I known that the Germans would not succeed
in producing an atomic bomb, I never would have
lifted a finger.

To *Newsweek* magazine, March 10, 1947, in regard to sending the famous letter to President Roosevelt about the
new possibility of constructing atom bombs. Einstein allegedly maintained that the development of nuclear energy
would have proceeded much the same even without his
intervention.

Through the release of atomic energy, our generation
has brought into the world the most revolutionary
force since prehistoric man's discovery of fire.

In a letter of support for the Emergency Committee of
Atomic Scientists, March 22, 1947. Einstein Archives 70-918

*[These military] tendencies . . . are something new
for America. They arose when, under the influence
of the two World Wars and the consequent concentration of all forces on a military goal, a predomi-
nantly military mentality developed, which with
the almost sudden victory became even more accentuated. The characteristic feature of this mentality is
that people place the importance of what Bertrand
Russell so tellingly terms "naked power" far above
all other factors which affect the relations between
peoples.

From "The Military Mentality," *American Scholar* 16, no. 3
(Summer 1947), 353–354. Reprinted in Rowe and Schulmann, *Einstein on Politics*, 477–479

It is characteristic of the military mentality that non-
human factors (atom bombs, strategic bases, weapons of all sorts, the possession of raw materials,
etc.) are held essential, while the human being, his
desires and thoughts—in short, the psychological
factors—are considered unimportant and secondary. . . . The individual is degraded . . . to "human
materiel."

Ibid.

The bombing of civilian centers was initiated by
the Germans and adopted by the Japanese. To it, the

Allies responded in kind—as it turned out, with greater effectiveness—and they were morally justified in doing so.

From "Einstein on the Atomic Bomb," part 2, an interview recorded by Raymond Swing, *Atlantic Monthly*, November 1947

Deterrence should be the only purpose of the stockpile of bombs. . . . To keep a stockpile of atomic bombs without promising not to initiate their use is to exploit possession of the bombs for political ends. . . . [Otherwise] atomic warfare will be hard to avoid.

Ibid.

The strength of the communist system of the East is that it has some of the character of a religion and inspires the emotions of a religion. Unless the force of peace, based on law, gathers behind it the force and zeal of religion, it can hardly hope to succeed. . . . There must be added that deep power of emotion that is a basic ingredient of religion.

Ibid.

It should not be forgotten that the atomic bomb was made in this country as a *preventive* measure; it was to head off its use by the Germans if they discovered it.

Ibid.

I am not saying the U.S. should not manufacture and stockpile the bomb, for I believe that it must do so; it must be able to deter another nation from making an atomic attack.

Ibid.

Since I do not foresee that atomic energy is to be a great boon for a long time, I have to say that for the present it is a menace. Perhaps it is good that it is so. It may intimidate the human race into bringing order into its international affairs, which, without the pressure of fear, it would not do.

The final words in ibid.

As long as there is man, there will be war.

To Philippe Halsman, 1947, quoted on p. 35 of *Time*, December 31, 1999, in its "Person of the Century" coverage of Einstein. Perhaps originally cited in Halsman's *Sight and Insight* (1972), which I have been unable to find.

All of us who are concerned about peace and the triumph of reason and justice must be keenly aware how small an influence reason and honest goodwill exert upon events in the political field.

From an address at Carnegie Hall in New York on receiving the One World Award, April 27, 1948. Published in *Out of My Later Years*; reprinted in *Ideas and Opinions*, 147

Where belief in the omnipotence of physical force gets the upper hand in political life, this force takes on a life of its own and proves stronger than the men who think to use force as a tool.

Ibid.

The proposed militarization of the nation not only immediately threatens us with war, it will also slowly but surely undermine the democratic spirit and the dignity of the individual in our land.

Ibid.

The victorious war against Nazi Germany and Japan has led to the unhealthy influence of our military men and of military attitudes which endangers the democratic institutions of our country and the peace of the world.

From a statement for Conference on "Pattern of Survival," New York, dated June 1, 1948. In Nathan and Norden, *Einstein on Peace*, 486. Bergreen Albert Einstein Collection, Vassar College, Box M2003-009, Folder 3.31. Einstein Archives 58-582

We scientists, whose tragic destiny it has been to help make the methods of annihilation ever more gruesome and more effective, must consider it our solemn and transcendent duty to do all in our power to prevent these weapons from being used for the brutal purpose for which they were invented.

Quoted in the *New York Times*, August 29, 1948

*It is true that the advances made in physics have made possible the application of scientific results for technical and military purposes involving great dangers. However, the responsibility lies with those who make use of the new tools and not those who contribute to the progress of knowledge: therefore, with the politicians and not with the scientists.

> Answer to a questionnaire from Milton James for the *Cheyney Record*, the student publication of Cheyney State Teachers College, a college for blacks in Pennsylvania, asking if the scientists who developed the bomb should be held morally responsible for its destructive outcome, October 7, 1948. Published February 1949. Corrected from my earlier version. See Nathan and Norden, *Einstein on Peace*, 501–502; Jerome and Taylor, *Einstein on Race and Racism*, 148. Einstein Archives 58-013 to 58-015

Creation of a United States of Europe is an economic and political necessity. Whether it would contribute to the stabilization of international peace is hardly predictable. I believe yes rather than no.

> In answer to the question of whether such an alliance would solve the problem of war. Ibid. (Corrected from my earlier version.)

I don't know [what weapons will be used in the Third World War]. But I can tell you what they'll use in the Fourth—rocks!

> From an interview, "Einstein at 70," with Alfred Werner, *Liberal Judaism* 16 (April–May 1949), 12. Einstein Archives 30-1104, p. 9 of typescript

So long as security is sought through national armament, no country is likely to renounce any weapon
that seems to promise it victory in the event of war.
In my opinion, security can be attained only by re-
nouncing all national military defense.

To Jacques Hadamard, December 29, 1949. Einstein Archives 12-064

*I have never done research having any bearing
upon the production of the atomic bomb. My sole
contribution in this field was that in 1905 I established the relationship between mass and energy, a
truth about the physical world of a very general nature, where possible connection to the military potential was completely foreign to my thoughts.

To A. J. Muste, January 23, 1950. Einstein Archives 60-631

If it [the effort to produce a hydrogen bomb] is successful, radioactive poisoning of the atmosphere
and hence annihilation of any life on Earth will have
been brought within the range of what is technically
possible.

From a contribution to Eleanor Roosevelt's television
program on the implications of the H-bomb, broadcast
February 12, 1950. Published in *Ideas and Opinions*,
159–161

Competitive armament is not a way to prevent
war. Every step in this direction brings us nearer to

catastrophe. . . . I repeat, *armament is no protection against war*, but leads inevitably to war.

> From a United Nations Radio interview, June 16, 1950, recorded in the study of Einstein's home in Princeton. Published in *Ideas and Opinions*, 161–163

Striving for peace and preparing for war are incompatible with each other. . . . Arms must be entrusted only to an international authority.

> Ibid.

The discovery of a nuclear chain reaction need not bring about the destruction of mankind any more than the discovery of matches.

> From a message for Canadian Education Week, March 1952. Einstein Archives 59-387

*The real ailment [lies] in . . . the belief that we must in peacetime so organize our whole life and work that in the event of war we would be sure of victory. This attitude gives rise to the belief that one's freedom and indeed one's existence are threatened by powerful enemies.

> From "Symptoms of Cultural Decay," *Bulletin of the Atomic Scientists* 8, no. 7 (October 1952), 217–218. See Rowe and Schulmann, *Einstein on Politics*, 48

The first atomic bomb destroyed more than the city of Hiroshima. It also exploded our inherited, outdated political ideas.

From a co-signed statement quoted in the *New York Times*,
June 12, 1953

*The Nuremberg Trial of the German war criminals
was tacitly based on the recognition of the principle:
criminal actions cannot be excused if committed on
government orders; conscience supersedes the authority of the law of the state.

> From "Human Rights," a message to the Chicago Deca-
> logue Society of Lawyers upon receiving its award for con-
> tributions to human rights. The message had been written
> just before December 5, 1953 (Einstein Archives 28-1012),
> and was translated and recorded before being played at the
> ceremony on February 20, 1954. See Rowe and Schulmann,
> *Einstein on Politics*, 497

*A.E.C. = Atomic Extermination Conspiracy.

> Jotted onto a copy of the pamphlet "Stop the Bomb: An At-
> tempt to the Reason of the American People," ca. April–
> June 1954. Einstein Archives 28-925. The pamphlet was
> sent to Einstein in the hope he would sign a petition to
> President Eisenhower urging him to discontinue testing
> hydrogen bombs in the Pacific. Einstein felt such an appeal
> was useless and only self-gratifying. See Schweber, "Ein-
> stein and Nuclear Weapons," in Galison, Holton, and
> Schweber, eds., *Einstein for the Twenty-first Century*, 91. Also
> mentioned in Johanna Fantova's journal, June 14, 1954.

*The only comfort which may be derived from the
development of atomic weapons is the hope that *this*
weapon may act as a deterrent and give impetus to
a movement to establish supranational safeguards.

Unfortunately, at the present time, the insanity of nationalism seems more powerful than ever before.

> To Seiei Shinohara, July 7, 1954. See Rowe and Schulmann, *Einstein on Politics*, 493. Einstein Archives 61-306

I made one great mistake in my life—when I signed that letter to President Roosevelt recommending that atom bombs be made; but there was some justi-fication—the danger that the Germans would make them!

> Written down by Linus Pauling in his diary after speaking with Einstein, November 16, 1954. Copied directly from the diary. The longer version quoted in the first two editions of this book (copied from secondary sources) is not in the diary. Although Einstein's letter was a warning about pos-sibilities and did not actually advocate the building of a bomb, he realized that it might ultimately lead to one. If Americans did not speed up their research in this area, Leo Szilard, who drafted the letter, and Einstein feared that Hit-ler might develop a bomb first and use it; that without one, the United States would not be able to retaliate in kind to protect itself; and that failure to write to Roosevelt might lead to a world ruled by a nuclear-armed Hitler. (Diary is at Valley Library, Oregon State University, Corvallis. Ava Helen and Linus Pauling Papers)

There was never even the slightest indication of any potential technological application.

> To Jules Isaac, February 28, 1955, refuting the idea that his special theory of relativity was responsible for atomic fis-sion and the atom bomb. Atomic fission, accomplished in December 1938 in Berlin by Otto Hahn and Fritz Strass-mann, was made possible by the discovery of the neutron

by James Chadwick in 1932; fission requires neutrons. Quoted in Nathan and Norden, *Einstein on Peace*, 623. Einstein Archives 59 1055

There lies before us, if we choose, continued progress in happiness, knowledge, and wisdom. Shall we, instead, choose death, because we cannot forget our quarrels? We appeal, as human beings, to human beings: Remember your humanity and forget the rest.

From the first paragraph of Einstein's last signed statement, regarding the development of weapons of mass destruction, drafted and signed by Bertrand Russell and signed by nine other scientists; signed by Einstein on April 11, 1955, one week before his death. The document has come to be known as the Russell-Einstein Manifesto, issued in London on July 9, 1955, after Einstein's death. Einstein Archives 33-211

We invite this Congress, and through it the scientists of the world and the general public, to subscribe to the following resolution: "In view of the fact that in any future world war nuclear weapons will certainly be employed, and that such weapons threaten the continued existence of mankind, we urge the Governments of the world to realize, and to acknowledge publicly, that their purpose cannot be furthered by a world war, and we urge them, consequently, to find peaceful means for the settlement of all matters of dispute between them."

Ibid.

On Politics, Patriotism, and Government

Einstein's politics can best be described by this passage from Rowe and Schulmann's *Einstein on Politics*, 458: "Einstein shunned ideologues, whether on the left or the right. By the same token, he had the deepest sympathy for all those who spoke out against tyranny and in favor of human freedom. Indeed, for the last twenty-five years of his life he was an indefatigable advocate of civil liberties and a staunch defender of those who put their lives in jeopardy to advance human rights."

*I have no intention of making a secret of my internationalist sentiments. How close I feel to a human being or a human organization depends only on how I judge their intentions and capabilities. The state, to which I as a citizen belong, has no place at all in my emotional life; I consider affiliation with a state to be a business matter, somewhat akin to one's relationship to life insurance.

A passage deleted by the editors from "My Opinion on the War," an essay for the Berliner Goethebund, published in 1916. According to Rowe and Schulmann, *Einstein on Politics*, 73, Einstein "could not resist pointing out that he was only reiterating Tolstoy's comparison of patriotism to a mental disorder." *CPAE*, Vol. 6, Doc. 20

*The impulsive behavior of contemporary man in political matters is enough to keep one's faith in determinism alive.

To Max Born, June 4, 1919. *CPAE*, Vol. 9, Doc. 56

I am convinced that what the next few years will bring will be far less difficult than the experiences of past years.

> An unprophetic statement about Germany's political and economic situation. Ibid.

When a group of people is possessed with collective insanity, one should act out against them; but hate and bitterness cannot consume a great and discerning people for the long term unless they are troubled themselves.

> To H. A. Lorentz, August 1, 1919, regarding the post–World War I Manifesto of 93, signed by ninety-three German intellectuals in defense of Germany. *CPAE*, Vol. 9, Doc. 80

I don't believe that humanity as such can change in essence, but I do believe it is possible and even necessary to put an end to anarchy in international relations, even though sacrifice of autonomy will be significant for individual states.

> To Hedwig Born, August 31, 1919. *CPAE*, Vol. 9, Doc. 97

On every cornfield, poisonous weeds can grow alongside the corn when conditions are right. I believe the conditions matter more than the soil.

> To Jean Perrin, September 27, 1919. *CPAE*, Vol. 9, Doc. 114

*I believe that international reconciliation would be advanced if young students and artists, in greater numbers than before, were to study in former enemy countries. Direct experience most effectively counteracts those disastrous ideologies which under the influence of the World War have been planted in many heads.

From "On the Contribution of Intellectuals to International Reconciliation," an essay for the German Social and Scientific Society of New York, ca. October 1920. *CPAE*, Vol. 7, Doc. 47

*Internationalism as I conceive it implies a rational relationship between countries, a sane union and understanding between nations, mutual cooperation, mutual advancement without interference with a country's customs or inner life.

From "On Internationalism," *New York Evening Post*, March 26, 1921. See Rowe and Schulmann, *Einstein on Politics*, 89; also Illy, *Albert Meets America*, 4

*I think the idea of a Bolshevik experiment should be excluded. Where they have had a Bolshevik experiment, as in Bavaria, the foolish reactionary ambitions have again become prevalent.

From an interview with Aldo Sorani in *Il Messagero*, October 26, 1921. *CPAE*, Vol. 12, Appendix G

In my opinion it is not right to bring politics into scientific matters, nor should individuals be held responsible for the government of the country to which they happen to belong.

To H. A. Lorentz, August 16, 1923. Quoted in French, *Einstein: A Centenary Volume*, 187. Einstein Archives 16-554

*Shudder to view this tragedy of human history where one murders out of fear that one will be murdered. It is the best, the most altruistic, who are tortured and killed because their political influence is feared—but not only in Russia. . . . [The rulers of Russia] will lose all sympathy if they cannot show through a great and courageous act of liberation that they do not need to rely on bloody terror to lend support to their political ideals.

On a collection of correspondence from inmates in the early Soviet gulag and affidavits of political persecution in the Soviet Union published by the International Committee of Political Prisoners in 1925. See Rowe and Schulmann, *Einstein on Politics*, 412–413. Einstein Archives 28-029

*Of course I am not a politician in the conventional sense of the word; few scholars are. At the same time I believe that no one should shirk the political task . . . of restoring *the unity between nations* that has been completely destroyed by the world war and seeing to it that a better and more genuine understanding among nations makes it impossible to repeat the dreadful catastrophe we have lived through.

From an interview in *Neue Zürcher Zeitung*, November 20, 1927. Einstein Archives 29-022

It is possible to be both. I look upon myself as a person. Nationalism is an infantile disease. It is the measles of mankind.

When asked if he considers himself a German or a Jew. From interview with G. S. Viereck, "What Life Means to Einstein," *Saturday Evening Post*, October 26, 1929. Quoted in Dukas and Hoffmann, *Albert Einstein, the Human Side*, 38; reprinted in Viereck, *Glimpses of the Great*, 449

*If there is anything that can give a layman in the sphere of economics the courage to express an opinion on the nature of the alarming economic difficulties of the present day, it is the hopeless confusion of opinions among the experts. . . . If we could somehow manage to prevent the purchasing power of the masses, measured in terms of goods, from sinking below a certain minimum, stoppages in the industrial cycle such as we are experiencing today would be rendered impossible.

From "Thoughts on the World Economic Crisis," ca. 1930, published in *Mein Weltbild* (1934). Einstein Archives 28-120. See Rowe and Schulmann, *Einstein on Politics*, 414–417.

My political ideal is democracy. Everyone should be respected as an individual, but no one idolized.

From "What I Believe," *Forum and Century* 84 (1930), 193–194. See Rowe and Schulmann, *Einstein on Politics*, 228

*I am convinced that degeneracy follows every autocratic system of violence, for violence inevitably attracts moral inferiors. Time has proved that illustrious tyrants are succeeded by scoundrels.

Ibid.

*I would never participate in [an ineffectual conference]. It would be like organizing a conference to stop volcanoes from erupting or to increase rain in the Sahara.

To Henri Barbusse, April 20, 1932. Einstein Archives 34-533

*Only in a free society is man able to create the inventions and cultural values which make life worthwhile to modern man.

From a speech in the Royal Albert Hall, "Science and Civilization," October 3, 1933. Published in 1934 as "Europe's Danger–Europe's Hope." Reprinted in Rowe and Schulmann, *Einstein on Politics*, 280. Einstein Archives 28-253

Nationalism is, in my opinion, nothing more than an idealistic rationalization for militarism and aggression.

From the second draft of the speech in the Royal Albert Hall, London, October 3, 1933. Quoted in Nathan and Norden, *Einstein on Peace*, 242. Einstein Archives 28-254

*If, as seems likely, the great-power democracies henceforth act with neutrality against Hitler's Germany, this breeding ground of disease will soon pose a grave moral and political danger for the rest of the world, not to speak of the unspeakable misery meted out to German Jews.

To Rabbi Stephen Wise, November 18, 1933. Einstein Archives 35-134

*No purpose is so high that unworthy methods in achieving it can be justified in my eyes. Violence sometimes may have cleared away obstructions quickly, but it has never proved itself creative.

From "Was Europe a Success?" *The Nation* 139, no. 3613 (October 3, 1934), 373. See Rowe and Schulmann, *Einstein on Politics*, 448

National loyalty is limiting; men must be taught to think in world terms. Every country will have to surrender a portion of its sovereignty through international cooperation. To avoid destruction, aggression must be sacrificed.

From an interview in *Survey Graphic* 24 (August 1935), 384, 413

Politics is a pendulum whose swings between anarchy and tyranny are fueled by perennially rejuvenated illusions.

Aphorism, 1937. Quoted in Dukas and Hoffmann, *Albert Einstein, the Human Side*, 38. Einstein Archives 28-388

Everything that is really great and inspiring is created by the individual who can labor in freedom.

From "Morals and Emotions," commencement address at Swarthmore College, June 6, 1938. Einstein Archives 29-083

*Scientists have an obligation to become politically active in the interest of free scientific research. They must have the courage . . . to enunciate with clarity their hard-won political and economic convictions.

To the Lincoln's Birthday Committee for Democracy and Intellectual Freedom, February 1939. Cited in Nathan and Norden, *Einstein on Peace*, 283

What distinguishes a true republic is not only the form of its government but also the deeply rooted feelings of equal justice for all and respect for every individual.

From a statement issued on Einstein's sixtieth birthday. *Science* 89, n.s. (1939), 242

*Excessive nationalism is a state of mind which is artificially induced by the prevalent obsession of nations that they must, at all times, be prepared for war. If the danger of war were eliminated, nationalism would soon disappear.

From "I Am an American," June 22, 1940. Einstein Archives 29-092. Reprinted in Rowe and Schulmann, *Einstein on Politics*, 470–472

There are times when the climate of the world is good for ethical things. Sometimes men trust one another and create good. At other times, it is not so.

From a conversation recorded by Algernon Black, Fall
1940. Einstein forbade the publication of this conversation.
Einstein Archives 54-834

When people live in a time of maladjustment, when
there is tension and disequilibrium, they become
unbalanced themselves and then may follow an unbalanced leader.

Ibid.

The greatest weakness of the democracies is economic fear.

Ibid.

*Laws alone cannot secure freedom of expression; in
order that every man presents his views without
penalty, there must be a spirit of tolerance in the entire population.

From Einstein's contribution to *Freedom, Its Meaning*, ed.
Ruth Nanda Anshen (1940). Einstein Archives 28-538

*What is a capitalist state? It is a state in which the
principal means of production, such as farmland,
real estate in the cities, the supply of water, gas, and
electricity, public transportation, as well as the larger
industrial plants are owned by a minority of the citizenry. Productivity is geared toward making a profit
for the owners rather than providing the population
with a uniform distribution of essential goods. . . .

A state can be characterized as "socialistic" when the principal means of production are owned collectively and are administered by individuals responsible to them and who are paid by the state.

> From "Is There Room for Individual Freedom in a Socialist State?" ca. July 1945. Einstein Archives 28-661. See Rowe and Schulmann, *Einstein on Politics*, 436

*My answer to this is that freedom, in any case, is only possible by constantly struggling for it. A citizenry that is politically indifferent will always end up enslaved no matter what form its constitution and legal institutions take. I am convinced, however, that in a state with a socialist economy the prospects are better for the average individual to achieve the maximum degree of freedom that is compatible with the well-being of the community.

> On the possibility of individual freedoms in a socialist society. Ibid.

*As for socialism, unless it is international to the extent of producing a world government which controls all military power, it might lead to wars even more easily than capitalism because it represents an even greater concentration of power.

> From "Einstein on the Atomic Bomb," part 1, an interview recorded by Raymond Swing, *Atlantic Monthly* 176, no. 5 (November 1945), 43–45. See Rowe and Schulmann, *Einstein on Politics*, 373–378

*If you occasionally hear my name mentioned in connection with political excursions, don't think that I spend much time on such things, as it would be sad to waste much energy on the meager soil of politics. From time to time, however, the moment arrives when I cannot help myself.

To Michele Besso, April 21, 1946. Einstein Archives 7-381

Democratic institutions and standards are the result of historic developments to an extent not always appreciated in the lands which enjoy them.

From "Einstein on the Atomic Bomb," part 2, an interview recorded by Raymond Swing, *Atlantic Monthly*, November 1947

To act intelligently in human affairs is only possible if an attempt is made to understand the thoughts, motives, and apprehensions of one's opponent so fully that one can see the world through his eyes.

From "Reply to the Soviet Scientists," December 1947, published in the *Bulletin of the Atomic Scientists* 4, no. 2 (February 1948), 35–37. Also see Rowe and Schulmann, *Einstein on Politics*, 393–397; and *Ideas and Opinions*, 140–146. Einstein Archives 28-795

*I also believe that capitalism or, we should say, the system of free enterprise will prove unable to check unemployment, which will become increasingly chronic because of technological progress, and unable to maintain a healthy balance between production and the purchasing power of the people.

Ibid.

*On the other hand, we should not make the mistake of blaming capitalism for all existing social and political evils, and of assuming that the very establishment of socialism would be able to cure all the social and political ills of humanity. The danger of such a belief lies, first, in the fact that it encourages fanatical intolerance on the part of all the "faithfuls" by making a possible social method into a type of church which brands all those who do not belong to it as traitors or as nasty evildoers. Once this stage has been reached, the ability to understand the convictions and actions of the "unfaithfuls" vanishes completely. You know, I am sure, from history how much unnecessary suffering such rigid beliefs have inflicted upon mankind.

Ibid.

Any government is in itself an evil insofar as it carries within it the tendency to deteriorate into tyranny.

Ibid.

*Socialism as such cannot be considered the solution to all social problems but merely as a framework within which such a solution is possible.

Ibid. See the similar statement for the *Cheyney Record* regarding democracy below.

*We are reminded by them that even the most perfect democratic institutions are no better than the people who are their agents.

For the Sacco and Vanzetti Memorial, 1947. Einstein Archives 28-770

We must learn the difficult lesson that the future of mankind will only be tolerable when our course, in world affairs as in all other matters, is based upon justice and law rather than the threat of naked power.

From a message for the Gandhi memorial service, February 11, 1948. Quoted in Nathan and Norden, *Einstein on Peace*, 468. Einstein Archives 5-151

*Democracy, taken in its narrower, purely political, sense suffers from the weakness that the bearers of the economic and political power have at their disposal powerful means to mold public opinion to serve their class interests. The democratic form of government in itself does not automatically solve the problems; it offers, however, a useful framework for their solution.

From an interview with Milton James of the *Cheyney Record*, the student publication of Cheyney State Teachers College in Pennsylvania, October 7, 1948. Corrected from my earlier version. See also the similar statement about socialism above. See Nathan and Norden, *Einstein on Peace*, 502. Einstein Archives 58-013 to 58-015

*Under the Russian system as it is today everything depends on the intentions and qualities of a few men in whose hands the whole power is concentrated. The position of the individual under that rule can be characterized thus: Considerable economic security at the expense of liberty and political rights.

Ibid.

*It is a happy fate to remain fascinated by one's work up to the last gasp. Otherwise we would suffer too much from the stupidity and madness of man as manifested mainly in politics.

To Michele Besso, July 24, 1949. Einstein Archives 7-386

Tito and Stalin's little dance shows socialism is not a path to gentleness.

To Otto Nathan, August 13, 1949. Bergreen Albert Einstein Collection, Vassar College, Box M2003-009, Folder 2.12. Einstein Archives 38-584

*It cannot be doubted that the achievements of the Soviet Regime are considerable in the fields of education, public health, social welfare, and economics, and that the people as a whole have greatly gained by these achievements.

To Sidney Hook, May 16, 1950. See, however, Rowe and Schulmann, *Einstein on Politics*, 456–457, for evidence that Einstein had "no sympathy at all for the dogmatic brand of Marxism [the Soviet regime] promulgated as official state doctrine." Einstein Archives 59-1018

I have never been a Communist. But if I were, I would not be ashamed of it.

To Lydia B. Hewes, July 10, 1950. Einstein Archives 59-984

Everyone who is in the business of dispensing reliable information today has the duty to enlighten the public. For even a conscientious person cannot reach reasonable political conclusions without trustworthy, factual information.

To Otto Nathan, November 5, 1950 (or perhaps May 11, 1950; dated 11-5-1950). Folder 2.14, Bergreen Albert Einstein Collection, Vassar College, Box M2003-009, Folder 2.14. Einstein Archives 38-586

I can see only the revolutionary way of non-cooperation in the sense of Gandhi's. Every intellectual who is called before one of the committees ought to refuse to testify; i.e., he must be prepared for jail and economic ruin . . . in the interest of the cultural welfare of his country.

To Brooklyn teacher William Frauenglass, who was called before the Senate's Internal Security Subcommittee (its equivalent of the House Un-American Activities Committee, or HUAC) hearings, May 16, 1953. Einstein Archives 41-112

Refusal to testify must be based on the assertion that it is shameful for a blameless citizen to submit to such an inquisition and that this kind of inquisition violates the spirit of the Constitution.

Ibid.

There is no such [anti-Communist] hysteria in the West European countries and there is no danger of their governments being overthrown by force or subversion, despite the fact that Communist parties are not persecuted or even ostracized.

To E. Lindsay, July 18, 1953. Einstein Archives 60-326

Eastern Europe would never have become prey to Russia if the Western powers had prevented German aggressive fascism under Hitler, which grave mistake made it necessary afterwards to beg Russia for help.

Ibid.

The fear of communism has led to practices that have become incomprehensible to the rest of civilized mankind and expose our country to ridicule. How long shall we tolerate politicians, hungry for power, who are trying to gain political advantage in such a way?

From "Human Rights," a message to the Chicago Decalogue Society of Lawyers upon receiving its award for contributions to human rights. The message had been written just before December 5, 1953 (Einstein Archives 28-1012), and was translated and recorded before being played at the ceremony on February 20, 1954. See Rowe and Schulmann, *Einstein on Politics*, 497

Party membership is a thing for which no citizen is obligated to give an accounting.

To C. Lamont, January 2, 1954. Einstein Archives 60-178

Yes, I'm an old revolutionary . . . politically I'm still a fire-spewing Vesuvius.

Quoted by Fantova, "Conversations with Einstein," February 9, 1954

*America is incomparably less endangered by its own Communists than by the hysterical hunt for the few Communists that are here. . . . Why should America be so much more endangered than England by the English Communists? Or is one to believe that the English are politically more naïve than the Americans so that they do not realize the danger they are in?

To Norman Thomas, March 10, 1954. Einstein Archives 61-549

The current [House Un-American Activities Committee and Senate Internal Security Subcommittee] investigations are an incomparably greater danger to our society than those few Communists in the country ever could be. These investigations have already undermined to a considerable extent the democratic character of our society.

To Felix Arnold, March 19, 1954. Einstein Archives 59-118

The Russians . . . want to give me a peace prize, but I have turned it down. That's all I need—to be called a Bolshevik here.

Quoted by Fantova, "Conversations with Einstein," April 2, 1954

In Plato's time, and even later, in Jefferson's time, it was still possible to reconcile democracy with a moral and intellectual aristocracy, while today democracy is based on a different principle—namely, that the other fellow is not better than I am.

On democracy and anti-intellectualism, in Niccolo Tucci's *New Yorker* profile of Einstein, November 22, 1954

A good government . . . is one which gives the citizen the maximum amount of liberty and political rights as is desirable in his own interest. On the other hand, the state has to provide for the citizen personal security and a certain amount of economic security. This situation necessitates a compromise between those two requirements which has to be found according to circumstances.

To Edward Shea, a Brooklyn police lieutenant, November 30, 1954. Einstein Archives 61-291

*I cannot rid myself of the thought that this, the last of my fatherlands, has invented for its own use a new kind of colonialism. . . . It achieves domination of other countries by investing American capital abroad, which makes those countries firmly dependent on the United States. Anyone who opposes this policy or its implications is treated as an enemy of the United States.

> In his last letter to Queen Elisabeth of Belgium, January 2, 1955, on America's postwar foreign policy. Einstein Archives 32-413

*Nothing astonishes me quite so much as the shortness of man's memory with regard to political developments. Yesterday the Nuremberg trials, today the all-out effort to rearm Germany.

> Ibid.

Political passions, once they have been fanned into flames, exact their victims.

> Final written words, from an unfinished draft of a radio address on occasion of the seventh anniversary of Israel's independence, probably April 10–2, 1955. See Rowe and Schulmann, *Einstein on Politics*, 507. See also Fantova, "Conversations with Einstein," April 10, 1955. Einstein Archives 28-1098

That is simple, my friend: because politics is more difficult than physics.

When asked why people could discover atoms but not the means to control them. Recalled in the *New York Times*, April 22, 1955, after Einstein's death.

One must divide one's time between politics and equations. But our equations are much more important to me, because politics is for the present, while our equations are for eternity.

Quoted by Ernst Straus in Seelig, *Helle Zeit, dunkle Zeit*, 71

On Race and Prejudice

Race is a fraud. All modern people are a conglomer-
ation of so many ethnic mixtures that no pure race
remains.

From an interview with G. S. Viereck, "What Life Means to
Einstein," *Saturday Evening Post*, October 26, 1929; reprinted in Viereck, *Glimpses of the Great*, 450

Insofar as we may at all claim that slavery has been
abolished today, we owe its abolition to the practical
consequences of science.

From "Science and Society," *Science*, Winter 1935–36. Reprinted in *Einstein on Humanism*, 11, and *Out of My Later
Years*, 135. Einstein Archives 28-324

*It is really a rather comforting thought that in India,
too, the all-too-human trait of knavery predomi-
nates. After all, it would be just too bad if this were
the privilege of the proud white race. I believe that
all creatures who can have young ones together are
very much the same.

To Max Born, ca. 1937–38. In Born, *Born-Einstein Letters*,
126. Born (p. 127) interprets the statement as Einstein's rejection of racial discrimination and national pride. Einstein
Archives 8-199

This country still has a heavy debt to discharge for
all the troubles and disabilities it has laid on the
Negro's shoulders. . . . To the Negro and his wonderful songs and choirs we are indebted for the finest

contribution in the realm of art which America has so far given the world.

At the dedication of the Wall of Fame at the 1939–40 World's Fair. Einstein Archives 28-527

*As regards conduct toward others, people would be truly democratic were it not for the still present dark shadow of racial prejudices, particularly toward Negroes. I believe that each individual must work within his or her circle to eradicate this shameful evil.

From "On Political Freedom in the U.S.A.," 1945. See Rowe and Schulmann, *Einstein on Politics*, 473. Einstein Archives 28-627

There is, however, a somber point in the social outlook of Americans. Their sense of equality and human dignity is mainly limited to men of white skin. . . . The more I feel like an American, the more this situation pains me.

From "A Message to My Adopted Country," *Pageant* 1, no. 12 (January 1946). Einstein supported the fledgling civil rights movement, perhaps also influenced by Paul Robeson, a black opera singer, former athlete, and early civil rights advocate, who was born in Princeton, and opera singer Marian Anderson, both of whom he befriended, as well as his cordial interest in and interactions with the black citizens of Princeton. See Rowe and Schulmann, *Einstein on Politics*, 474. See also, in general, Jerome and Taylor, *Einstein on Race and Racism*.

The modern prejudice against Negroes is the result of the desire to maintain this unworthy condition.

Ibid., 475, referring to the "fatal misconception" that blacks "are not our equals in intelligence, sense of responsibility, reliability," and further stating that these are ancient prejudices against slaves, both black and white

We must recognize what in our accepted tradition is damaging to our fate and dignity—and shape our lives accordingly.

Ibid.

*Your ancestors dragged these black people from their homes by force; and in the white man's quest for wealth and an easy life they have been ruthlessly suppressed and exploited, degraded into slavery. The modern prejudice against Negroes is the result of the desire to maintain this unworthy condition.

Ibid.

I believe that whoever tries to think things through honestly will soon recognize how unworthy and even fatal is the traditional bias against Negroes. . . . What can the man of good will do to combat this deeply rooted prejudice? He must have the courage to set an example by words and deed, and must watch lest his children become influenced by racial bias.

Ibid., 475–476

*The worst disease from which the society of our nation suffers is . . . the treatment of the Negro. . . . Everyone who freshly learns of this state of affairs at a maturer age feels not only the injustice, but the scorn of the principle of the Fathers who founded the United States that "all men are created equal."

> From a letter to the National Urban League convention, September 16, 1946. See Jerome and Taylor, *Einstein on Race and Racism*, 144–145. Einstein Archives 54-543

*One can hardly believe that a reasonable man can cling so tenaciously to such prejudice, and there is sure to come a time in which school-children in their history lessons will laugh about the fact that something like this did once exist.

> Ibid., 145

*We must strive . . . that minorities be protected against economic and political discrimination as well as against attack by libelous writings and against the poisoning of youth in the schools. These endeavors are important, but not as important as the intellectual and moral enlightenment of the people.

> Ibid.

*To insure such protection [against acts of violence] is one of the most urgent tasks for our generation. A way always exists to overcome legal obstacles whenever there is a determined will in the service of such a just cause.

> To President Harry Truman, September 1946, in support of the American Crusade to End Lynching, which Einstein co-chaired with Paul Robeson. Quoted in the *New York Times*, September 23, 1946. See Jerome and Taylor, *Einstein on Race and Racism*, 143. Einstein Archives 57-103

*On reading the White article one is struck with the deep meaning of the saying: There is only one road to true human greatness—the road through suffering. If the suffering springs from the blindness and dullness of a tradition-bound society, it usually degrades the weak to a state of blind hate.

> Letter to the editor of the *Saturday Review of Literature*, November 11, 1947, commenting on Walter White's article, "Why I Remain a Negro" (*SRL*, October 1947). White chose to identify as a black and worked in the civil rights movement even though he was fair-skinned and blond enough to pass as white. See Jerome and Taylor, *Einstein on Race and Racism*, 147. Einstein Archives 28-768

*[Walter White] has allowed us to accompany him on the painful road to human greatness by giving us a simple biographical story which is irresistible in its convincing power.

> Ibid., 148

*The more cruel the wrong that men commit against an individual or a people, the deeper their hatred and contempt for their victim. Conceit and false pride on the part of a nation prevent the rise of remorse for its crime. Those who have had no part in the crime, however, have no sympathy for the sufferings of the innocent victims of persecution and no awareness of human solidarity.

> From statement for the Monument to the Martyred Jews of the Warsaw Ghetto, April 19, 1948. See Rowe and Schulmann, *Einstein on Politics*, 349. Einstein Archives 28-815

*Race prejudice is a part of a tradition which—conditioned by history—is uncritically handed down from one generation to another. The only remedy is enlightenment and education. This is a slow and painstaking process in which all right-thinking people should take part.

> In answer to an interview by Milton James for the *Cheyney Record*, the student newspaper of Cheyney State Teachers College, a college for blacks in Pennsylvania, asking if U.S. racial prejudice was a symptom of a worldwide conflict, October 7, 1948, published February 1949. Reprinted thus in Nathan and Norden, *Einstein on Peace*, 502: "Race prejudice has unfortunately become an American tradition which is uncritically handed down from one generation to the next. The only remedies are enlightenment and education. This is a slow and pain-staking process in which all right-thinking people should take part." Note that "American" was not in the handwritten facsimile in Kaller's autographs catalog and in the newspaper article but was added in *Einstein on Peace*. The copy (perhaps a draft) in the autographs catalog, addressed to Milton M. James of the *Record*, has it thus: "Prejudice is part of a tradition which—determined through history—is handed down uncritically from

generation to generation. One can achieve liberation from prejudice only by enlightenment and education. This is a slow and painstaking purifying process in which every concerned person has to participate." Einstein Archives 58-013 to 58-015

On Religion, God, and Philosophy

The way he often explained it, Einstein's "religion" was an attitude of cosmic awe and wonder and a devout humility before the harmony of nature, rather than a belief in a personal God who is able to control the lives of individuals. He referred to this belief as "cosmic religion." It is incompatible with the doctrines of all theistic religions in its denial of a personal God who punishes the wicked and rewards the righteous. Einstein was an admirer of Spinoza, the seventeenth-century Dutch Jewish rationalist philosopher, whom the German Romantic poet Novalis called a "God-intoxicated man." Because of Einstein's frequent reference to God, one might be tempted to think of him in this way as well. For a thorough discussion on Einstein's religion, see Jammer, *Einstein and Religion*. See also "Mysticism" in the "Miscellaneous" section.

Why do you write to me, "God should punish the English"? I have no close connection to either one or the other. I see only with deep regret that God punishes so many of his children for their numerous stupidities, for which only he himself can be held responsible; in my opinion, only his nonexistence could excuse him.

To Edgar Meyer, a Swiss colleague, January 2, 1915. *CPAE*, Vol. 8, Doc. 44

*Why [use] so many words when I can say all in a ... sentence appropriate for a Jew: Honor your

Master Jesus Christ not only in words and hymn, but above all by your deeds.

> From "My Opinion on the War," in the Berliner Goethe-bund's *The Land of Goethe 1914/1916*, published in 1916. *CPAE*, Vol. 6, Doc. 20

Upon reading books on philosophy, I learned that I stood there like a blind man in front of a painting. I can grasp only the inductive method . . . the works of speculative philosophy are beyond my reach.

> To Eduard Hartmann, April 27, 1917. *CPAE*, Vol. 8, Doc. 330

The suprapersonal content conveyed by religion, primitive in form though it is, is more valuable, I am convinced, than Haeckel's materialism. I believe that even nowadays, eliminating the sacred traditions would still mean spiritual and moral impoverishment—as gross and ugly as the attitude and actions of the clergy may be in many respects.

> To Georg Count von Arco, January 14, 1920, in declining to be identified as a Monist. Ernst Haeckel was a relentless fighter against prominent traditional religious doctrines, but he also alienated many freethinkers with his views on eugenics, race, and his conservative political agenda. His brutal social ethic influenced the Nazis. *CPAE*, Vol. 9, Doc. 260

In every true searcher of Nature there is a kind of religious reverence, for he finds it impossible to imagine that he is the first to have thought out the exceedingly delicate threads that connect his perceptions.

1920. Quoted by Moszkowski, *Conversations with Einstein*, 46

Since our inner experiences consist of reproductions and combinations of sensory impressions, the concept of a soul without a body seems to me to be empty and devoid of meaning.

To Viennese poet Lili Halpern-Neuda, February 5, 1921. Quoted in Dukas and Hoffmann, *Albert Einstein, the Human Side*, 40. *CPAE*, Vol. 12, Doc. 7

The meaning of the word "truth" varies according to whether we deal with a fact of experience, a mathematical proposition, or a scientific theory. "Religious truth" conveys nothing clear to me at all.

In answer to the question, Do scientific and religious truths come from different points of view? December 14, 1922, posed by interviewers for the Japanese magazine *Kaizo* 5, no. 2 (1923), 197. Reprinted in *Ideas and Opinions*, 261–262

Scientific research can reduce superstition by encouraging people to think and view things in terms of cause and effect. It is certain that a conviction akin to a religious feeling, of the rationality or intelligibility of the world lies behind all scientific work of a higher order.

In answer to the question, Can scientific discovery enhance religious belief and repudiate superstition, since religious feelings can give impetus to scientific discovery? Ibid.

My comprehension of God comes from the deeply felt conviction of a superior intelligence that reveals itself in the knowable world. In common terms, one can describe it as "pantheistic" (Spinoza).

In answer to the question, What is your understanding of God? Ibid.

I can look at doctrinaire traditions only with a historical and psychological perspective; they have no other significance for me.

In answer to the question, What is your opinion regarding a "savior"? Ibid.

I want to know how God created this world. I am not interested in this or that phenomenon, in the spectrum of this or that element. I want to know his thoughts. The rest are details.

Recalled by his Berlin student Esther Salaman, 1925, in Salaman, "A Talk with Einstein," *Listener* 54 (1955), 370–371

Try and penetrate with our limited means the secrets of nature and you will find that, behind all the discernible concatenations, there remains something subtle, intangible, and inexplicable. Veneration for this force beyond anything that we can comprehend is my religion. To that extent I am, in point of fact, religious.

From a dinner conversation between Einstein and the German critic Alfred Kerr, recorded by Count Harry Kessler in

his diary, *The Diary of a Cosmopolitan* (1971), June 14, 1927.
Also quoted in Brian, *Einstein, a Life*, 161

I cannot conceive of a personal God who would directly influence the actions of individuals. . . . My religiosity consists of a humble admiration of the infinitely superior spirit that reveals itself in the little that we can comprehend of the knowable world. That deeply emotional conviction of the presence of a superior reasoning power, which is revealed in the incomprehensible universe, forms my idea of God.

To M. Schayer, August 1, 1927. Quoted in Dukas and Hoffmann, *Albert Einstein, the Human Side*, 66, and in Einstein's *New York Times* obituary, April 19, 1955. Einstein Archives 48-380

I often read the Bible, but its original text has remained beyond my reach.

To H. Friedmann, March 18, 1929, regarding his lack of knowledge of the Hebrew language. Quoted in Pais, *Subtle Is the Lord*, 38. Einstein Archives 30-405

I believe in Spinoza's God, Who reveals Himself in the lawful harmony of the world, not in a God who concerns himself with the fate and the doings of mankind.

In answer to Rabbi Herbert S. Goldstein's telegram, published in the *New York Times*, April 25, 1929. (Spinoza reasoned that God and the material world are indistinguishable; the better one understands how the universe works, the closer one comes to God.) Goldstein felt this answer was

evidence that Einstein was not an atheist. See Rowe and
Schulmann, *Einstein on Politics*, 17. Einstein Archives 33-272

Everything is determined . . . by forces over which
we have no control. It is determined for the insect as
well as for the star. Human beings, vegetables, or
cosmic dust—we all dance to a mysterious tune, in-
toned in the distance by an invisible piper.

From an interview with G. S. Viereck, "What Life Means to
Einstein," *Saturday Evening Post*, October 26, 1929; reprinted in Viereck, *Glimpses of the Great*, 452

No one can read the Gospels without feeling the actual presence of Jesus. His personality pulsates in
every word. No myth is filled with such life.

In answer to the question, "Do you accept the historical
Jesus?" Ibid. Reprinted in Viereck, *Glimpses of the Great*,
448. Quoted in Brian, *Einstein, a Life*, 277 (a slightly different version is found on his p. 186). According to Brian, *Einstein, a Life*, 278, Einstein reportedly considered this interview an accurate representation of his views. Others regard
it with extreme caution.

I am not an atheist. I do not know if I can define myself as a pantheist. The problem involved is too vast
for our limited minds.

In answer to the question, "Do you believe in God?" Ibid.
Reprinted in Viereck, *Glimpses of the Great*, 447

*I do not believe that a moral philosophy can ever
be founded on a scientific basis. . . . The valuation of

life and all its nobler expressions can only come out of the soul's yearning toward its own destiny. Every attempt to reduce ethics to scientific formulas must fail. . . . On the other hand, it is undoubtedly true that scientific study of the higher kinds and general interest in scientific theory have great value in leading men toward a worthier valuation of the things of the spirit.

From "Science and God: A Dialogue," an exchange of opinions between Einstein, James Murphy, and J.W.N. Sullivan in *Forum and Century* 83 (June 1930), 373–379

I am of the opinion that all the finer speculations in the realm of science spring from a deep religious feeling. . . . I also believe that this kind of religiousness . . . is the only creative religious activity of our time.

Ibid.

There are two different conceptions about the nature of the universe: (1) the world as a unity dependent on humanity; (2) the world as a reality independent of the human factor.

From a conversation with Indian mystic, poet, and musician Rabindranath Tagore, summer 1930. Published in the *New York Times Magazine*, August 10, 1930

I cannot prove scientifically that Truth must be conceived as a truth that is valid independent of humanity, but I firmly believe it. . . . If there is a reality independent of man, there is also a truth relative to this reality.... The problem begins with whether Truth is independent of our consciousness.... For instance, if nobody is in this house, that table remains where it is.

Ibid.

The man who is thoroughly convinced of the universal operation of the law of causation cannot for a moment entertain the idea of a being who interferes in the course of events.... He has no use for the religion of fear and equally little for social or moral religion. A God who rewards and punishes is inconceivable to him for the simple reason that a man's actions are determined by necessity, external and internal, so that in God's eyes he cannot be responsible, any more than an inanimate object is responsible for the motions it undergoes.... A man's ethical behavior should be based effectively on sympathy, education, and social relationships; no religious basis is necessary. Man would indeed be in a poor way if he had to be restrained by fear of punishment and hope of reward after death.

From "Religion and Science," *New York Times Magazine*, November 9, 1930, 1–4. Reprinted in *Ideas and Opinions*, 36–40. See also *Berliner Tageblatt*, November 11, 1930. This version is taken from *Ideas and Opinions*, 36–40, whose translators chose not to follow the *New York Times* version.

Everything the human race has done and thought is concerned with the satisfaction of deeply felt needs and the assuagement of pain. One has to keep this constantly in mind if one wishes to understand religious movements and their development. Feeling and longing are the motive force behind all human endeavor and human creation, in however exalted a guise the latter may present themselves to us.

Ibid.

*The beginnings of cosmic religious feeling already appear at an early stage of development, e.g., in many of the Psalms of David and in some of the Prophets. Buddhism, as we have learned especially from the wonderful writings of Schopenhauer, contains a much stronger element of this.

Ibid.

It is very difficult to elucidate this [cosmic religious] feeling to anyone who is entirely without it. . . . The religious geniuses of all ages have been distinguished by this kind of religious feeling, which knows no dogma and no God conceived in man's image; so that there can be no church whose central teachings are based on it. . . . In my view, it is the most important function of art and science to awaken this feeling and keep it alive in those who are receptive to it.

On "cosmic religion," a worship of the harmony and beauties of nature that became the common faith of many physicists. Ibid.

I will call it the cosmic religious sense. This is hard to make clear to those who do not experience it, since it does not involve an anthropomorphic idea of God; the individual feels the vanity of human desires and aims, and the nobility and marvelous order which are revealed in nature and in the world of thought.

Ibid.

I assert that the cosmic religious experience is the strongest and the noblest driving force behind scientific research.

Ibid.

*I cannot imagine a God who rewards and punishes the objects of his creation, whose purposes are modeled after our own—a God, in short, who is but a reflection of human frailty. Neither can I believe that the individual survives the death of his body, although feeble souls harbor such thoughts through fear or ridiculous egotism. It is enough for me to contemplate the mystery of conscious life perpetuating itself through all eternity, to reflect upon the marvelous structure of the universe which we can dimly perceive, and to try humbly to comprehend even an infinitesimal part of the intelligence manifested in nature.

From "What I Believe," *Forum and Century* 84 (1930), 193–194. Reprinted in Rowe and Schulmann, *Einstein on Politics,*

229–230. Variously translated and used elsewhere, including in earlier editions of this book.

The most beautiful thing we can experience is the mysterious. It is the source of all true art and science. He to whom this emotion is a stranger, who can no longer pause to wonder and stand rapt in awe, is as good as dead: his eyes are closed. This insight into the mystery of life, coupled though it be with fear, has also given rise to religion. To know that what is impenetrable to us really exists, manifesting itself as the highest wisdom and the most radiant beauty which our dull faculties can comprehend only in their most primitive forms—this knowledge, this feeling, is at the center of true religiousness. In this sense, and in this sense only, I belong in the ranks of devoutly religious men.

Ibid.

*We must never forget that the most courageous fighters against militarism come from a religious group, from among the Quakers.

To Henri Barbusse, June 17, 1932. Einstein Archives 34-546

Philosophy is like a mother who gave birth to and endowed all the other sciences. Therefore, one should not scorn her in her nakedness and poverty, but should hope, rather, that part of her Don Quixote ideal will live on in her children so that they do not sink into philistinism.

To Bruno Winawer, September 8, 1932. Quoted in Dukas and Hoffmann, *Albert Einstein, the Human Side*, 106. Einstein Archives 52-267

Our actions should be based on the ever-present awareness that human beings in their thinking, feeling, and acting are not free but are just as causally bound as the stars in their motion.

From a statement to the Spinoza Society of America, September 22, 1932. Einstein Archives 33-291

If one purges all subsequent additions from the original teachings of the Prophets and Christianity, especially those of the priests, one is left with a doctrine that is capable of curing all the social ills of humankind.

From a statement for the Romanian Jewish journal *Renasterea Noastra*, January 1933. Published in *Mein Weltbild*; reprinted in *Ideas and Opinions*, 184–185

Organized religion may regain some of the respect it lost in the last war if it dedicates itself to mobilizing the good-will and energy of its followers against the rising tide of illiberalism.

Quoted in the *New York Times*, April 30, 1934. Broadcasted message for Brotherhood Day. Also quoted in Pais, *Einstein Lived Here*, 205

You will hardly find one among the profounder sort of scientific minds without a religious feeling of his own. But it is different from the religiosity of the naïve man. For the latter, God is a being from whose care one hopes to benefit and whose punishment one fears; a sublimation of a feeling similar to that of a child for its father.

From "The Religious Spirit of Science." Published in *Mein Weltbild* (1934), 18; reprinted in *Ideas and Opinions*, 40

The scientist is possessed by a sense of universal causation. . . . His religious feeling takes the form of a rapturous amazement at the harmony of natural law, which reveals an intelligence of such superiority that, compared with it, all the systematic thinking and acting of human beings is an utterly insignificant reflection. . . . It is beyond question closely akin to that which has possessed the religious geniuses of all ages.

Ibid.

What is the meaning of human life, or for that matter, of the life of any creature? To know an answer to this question means to be religious. You ask: Does it make any sense, then, to pose this question? I answer: The man who regards his own life and that of his fellow creatures as meaningless is not merely unhappy but hardly fit for life.

Published in *Mein Weltbild* (1934), 10; reprinted as "The Meaning of Life" in *Ideas and Opinions*, 11

Everyone has been given an endowment that he must strive to develop in the service of mankind. This cannot be brought to completion through the threat of a God who will punish man for sin, but only by challenging the best in human nature.

From an interview in *Survey Graphic* 24 (August 1935), 384, 413

Whatever there is of God and goodness in the universe, it must work itself out and express itself through us. We cannot stand aside and let God do it.

From a conversation recorded by Algernon Black, Fall 1940. Einstein forbade the publication of this conversation. Einstein Archives 54-834

To [the sphere of religion] belongs the faith in the possibility that the regulations valid for the world of existence are rational, that is, comprehensible to reason. I cannot conceive of a genuine scientist without that profound faith.

From "Science, Philosophy, and Religion," a written contribution to a symposium held in New York in 1940 on how science, philosophy, and religion advance the cause of American democracy; published in 1941 by the Conference on Science, Philosophy and Religion in Their Relation to the Democratic Way of Life. Reprinted in *Ideas and Opinions* as "Science and Religion," 44–47. Einstein Archives 28-523

A religious person is devout in the sense that he has no doubt about the significance of those superpersonal objects and goals that neither require nor are capable of rational foundation.

Ibid. See *Ideas and Opinions*, 45

Science without religion is lame, religion without science is blind.

Ibid. See *Ideas and Opinions*, 46. This may be a play on Kant's "Notion without intuition is empty, intuition without notion is blind"; Einstein was not always totally original. Some scientists, perhaps many, disagree with Einstein's sentiment. (See, for example, Dyson, "Writing a Foreword for Alice Calaprice's New Einstein Book," 491–502.)

The main source of the present-day conflicts between the spheres of religion and science lies in the concept of a personal God.

Ibid. See *Ideas and Opinions*, 47

The highest principles for our aspirations and judgments are given to us in the Judeo-Christian religious tradition. It is a very high goal that we, with our very weak powers, can reach only very inadequately, but which gives a sure foundation to our aspirations and values. . . . There is no room in this for the deification of a nation, of a class, let alone of an individual. Are we not all children of one Father, as it is said in religious language?

Ibid. See *Ideas and Opinions*, 43

It is only to the individual that a soul is given.

Ibid. That is, not to a class or nation.

In their struggle for the ethical good, teachers of religion must have the stature to give up the doctrine of a personal God, that is, give up that source of fear and hope which in the past placed such vast power in the hands of priests.

Ibid., 48

Whoever has undergone the intense experience of successful advances made in this domain [science] is moved by profound reverence for the rationality made manifest in existence.

Ibid., 49

The further the spiritual evolution of mankind advances, the more certain it seems to me that the path to genuine religiosity does not lie through the fear of life, and the fear of death, and blind faith, but through striving after rational knowledge.

Ibid.

In view of such harmony in the cosmos which I, with my limited human mind, am able to recognize, there are yet people who say there is no God. But what makes me really angry is that they quote me for support of such views.

Said to German anti-Nazi diplomat and author Hubertus
zu Löwenstein around 1941. Quoted in his book, *Towards
the Further Shore* (London, 1968), 156. With this remark,
Einstein dissociates himself from atheism; see Jammer, *Einstein and Religion*, 97

Then there are the fanatical atheists whose intolerance is the same as that of the religious fanatics, and
it springs from the same source. . . . They are creatures who can't hear the music of the spheres.

To an unidentified person, August 7, 1941, on the reaction
to his symposium contribution, "Science, Philosophy, and
Religion" (1940). To many readers, Einstein's denial of a
"personal" God meant a total denial of God, because "there
is no other God but a personal God." See discussion in
Jammer, *Einstein and Religion*, 92–108. Einstein Archives
54-927

It is quite possible that we can do greater things than
Jesus, for what is written in the Bible about him is
poetically embellished.

Quoted in W. Hermanns, "A Talk with Einstein," October
1943. Einstein Archives 55-285

No idea is conceived in our mind independent of
our five senses [i.e., no idea is divinely inspired].

Ibid.

I would not think that philosophy and reason themselves will be man's guide in the foreseeable future;

however, they will remain the most beautiful sanctuary they have always been for the select few.

To Benedetto Croce, June 7, 1944. Quoted in Pais, *Einstein Lived Here*, 122. Einstein Archives 34-075

Thus I came . . . to a deep religiosity, which, however, found an abrupt ending at the age of 12. Through the reading of popular scientific books I soon reached the conviction that much in the stories of the Bible could not be true. . . . Suspicion against every kind of authority grew out of this experience . . . an attitude which has never again left me.

Written in 1946 for "Autobiographical Notes," 3–5

Out yonder there was this huge world, which exists independently of us human beings and which stands before us like a great, eternal riddle, at least partially accessible to our inspection and thinking. The contemplation of this world beckoned like a liberation, and I soon noticed that many a man I had learned to esteem and admire had found inner freedom and security in devoted occupation with it.

Ibid., 5

My views are near those of Spinoza: admiration for the beauty and belief in the logical simplicity of the order and harmony that we can grasp humbly and only imperfectly. I believe that we have to content ourselves with our imperfect knowledge and understanding and treat values and moral obligations as purely human problems.

> To Marvin Magalaner, April 26, 1947. Quoted in Hoffmann, *Albert Einstein: Creator and Rebel*, 95. Einstein Archives 58-461

It is this . . . symbolic content of the religious traditions which is likely to come into conflict with science. . . . Thus it is of vital importance for the preservation of true religion that such conflicts be avoided when they arise from subjects which, in fact, are not really essential for the pursuit of religious aims.

> From a statement to the Liberal Ministers Club, New York City. Published in the *Christian Register*, June 1948; reprinted as "Religion and Science: Irreconcilable?" in *Ideas and Opinions*, 49–52

While it is true that scientific results are entirely independent of religious or moral considerations, those individuals to whom we owe the great creative achievements in science were all imbued with the truly religious conviction that this universe of ours is something perfect and is responsive to the rational striving for knowledge.

> Ibid.

A human being is a part of the whole, called by us "Universe," a part limited in time and space. He experiences himself, his thoughts and feelings as something separate from the rest—a kind of optical delusion of his consciousness. The striving to free oneself from this delusion is the one issue of true religion. Not to nourish it but to try to overcome it is the way to reach the attainable measure of peace of mind.

> To Robert Marcus, a distraught father who asked Einstein for some comforting words after the death of his young son, February 12, 1950. In Calaprice, *Dear Professor Einstein*, 184. Einstein Archives 60-424

My position concerning God is that of an agnostic. I am convinced that a vivid consciousness of the primary importance of moral principles for the betterment and ennoblement of life does not need the idea of a law-giver, especially a law-giver who works on the basis of reward and punishment.

> To M. Berkowitz, October 25, 1950. Einstein Archives 59-215

I have found no better expression than "religious" for confidence in the rational nature of reality, insofar as it is accessible to human reason. Whenever this feeling is absent, science degenerates into uninspired empiricism.

> To Maurice Solovine, January 1, 1951. Published in *Letters to Solovine*, 119. Einstein Archives 21-474

Mere unbelief in a personal God is no philosophy at all.

To V. T. Aaltonen, May 7, 1952, on his opinion that belief in a personal God is better than atheism. Einstein Archives 59-059

My feeling is religious insofar as I am imbued with the consciousness of the insufficiency of the human mind to understand more deeply the harmony of the universe which we try to formulate as "laws of nature."

To Beatrice Frohlich, December 17, 1952. Einstein Archives 59-797

The idea of a personal God is quite alien to me and seems even naïve.

Ibid.

To assume the existence of an unperceivable being . . . does not facilitate understanding the orderliness we find in the perceivable world.

To D. Albaugh, an Iowa student who asked, "What is God?" July 21, 1953. Einstein Archives 59-085

I do not believe in the immortality of the individual, and I consider ethics to be an exclusively human concern with no superhuman authority behind it.

To A. Nickerson, a Baptist pastor, July 1953. Quoted in Dukas and Hoffmann, *Albert Einstein, the Human Side*, 39. Einstein Archives 36-553

*The word God is for me nothing more than the expression and product of human weaknesses, and the Bible a collection of honorable but still primitive, rather childish legends. No interpretation, no matter how elegant, can change this [for me].

> To philosopher Eric Gutkind, January 3, 1954. See more in the section "On Jews, Israel, Judaism, and Zionism." The half-page handwritten letter sold for £170,000 ($404,000) at a Bloomsbury auction in London on May 15, 2008, a record for a single Einstein letter and twenty-five times the presale estimate. *New York Times*, May 17, 2008. Einstein Archives 33-337

If God created the world, his primary concern was certainly not to make its understanding easy for us.

> To David Bohm, February 10, 1954. Einstein Archives 8-041

I consider the Society of Friends the religious community that has the highest moral standards. As far as I know, they have never made evil compromises and are always guided by their conscience. In international life, especially, their influence seems to me very beneficial and effective.

> To A. Chapple, Australia, February 23, 1954. Quoted in Nathan and Norden, *Einstein on Peace*, 511. Einstein Archives 59-405

I do not believe in a personal God and I have never denied this but have expressed it clearly. If something is in me that can be called religious, then it is the unbounded admiration for the structure of the world so far as science can reveal it.

To an admirer who questioned him about his religious beliefs, March 22, 1954. Quoted in Dukas and Hoffmann, *Albert Einstein, the Human Side*, 43. Einstein Archives 39-525

I don't try to imagine a God; it suffices to stand in awe of the structure of the world, insofar as it allows our inadequate senses to appreciate it.

To S. Flesch, April 16, 1954. Einstein Archives 30-1154

A man's moral worth is not measured by what his religious beliefs are, but rather by what emotional impulses he has received from Nature.

To Sister Margrit Goehner, February 1955. Einstein Archives 59-830

Isn't all of philosophy like writing in honey? It looks wonderful at first sight, but when you look again it is all gone. Only the smear is left.

As recalled by Rosenthal-Schneider, *Reality and Scientific Truth*, 90

*As long as you pray to God and ask him for something, you are not a religious man.

In a conversation with Leo Szilard, date unknown, cited in Spencer R. Weart and Gertrud Weiss Szilard, eds., *Leo Szilard: His Version of the Facts* (Cambridge, Mass.: MIT Press, 1978), 12. Quoted slightly differently in Jammer,

Einstein and Religion, 149, where the words are probably wrongly sourced as having been *written* to Szilard; after a thorough search, we were unable to find a letter in the archives containing these words. Thanks to Szilard specialist Gene Dannen for clearing up this long-standing puzzle.

What really interests me is whether God could have created the world any differently; in other words, whether the requirement of logical simplicity admits a margin of freedom.

Quoted by Einstein's assistant Ernst Straus, on the question of whether God had any choice in the design of the world, in Seelig, *Helle Zeit, dunkle Zeit*, 72. Believers in intelligent design creation interpret this statement as support for an intelligent designer, not grasping that Einstein was using a metaphor while puzzling if it is possible to build more than one universe that is logically consistent.

We know nothing about it all [God, the world]. All our knowledge is but the knowledge of schoolchildren. Possibly we shall know a little more than we do now. But the real nature of things, that we shall never know, never.

From an interview with Chaim Tchernowitz, *The Sentinel*, date unknown

Papagoyim.

Einstein's name for the followers of the Catholic Church—i.e., the *goyim* (non-Jews) who follow the *papa*, or pope. The German for parrot is *Papagei*, and, according to archivist Barbara Wolff, Einstein is surely playing on that word to mean that the followers of the pope parrot what he says. It is also evocative of the plumage-clad Papageno and Papagena in Mozart's *The Magic Flute*. Thanks to Einstein scholar John Stachel, a former editor of *CPAE*, for this gem, passed on to him by Einstein's secretary, Helen Dukas.

On Science and Scientists, Mathematics, and Technology

Einstein is most famous for his theory of relativity. In his first papers he referred to it as the "relativity principle." The term "theory of relativity" was first used by Max Planck in 1906 to describe the Lorentz-Einstein equations of motion for the electron, and it was finally adopted by Einstein in 1907 in his reply to an article by Paul Ehrenfest, who had also used Planck's term. But Einstein continued to use "relativity principle" in the titles of articles for several years, since a "principle" is not a "theory" but something that is borne in mind when formulating a theory. In 1915, Einstein began to refer to the 1905 theory, having to do with space and time, as the "special theory," to distinguish it from his new theory of gravitation, the "general theory." See Stachel et al., *Einstein's Miraculous Year*, 101–102; and Fölsing, *Albert Einstein*, 208–210

Examples of a similar kind, and the failure of attempts to detect a motion of the Earth relative to the "light medium," lead to the conjecture that not only in mechanics, but in electrodynamics as well, the phenomena do not have any properties corresponding to the concept of absolute rest, but that in all coordinate systems in which the mechanical equations are valid, the same electrodynamic and optical laws are also valid, as has already been shown for quantities of the first order.

This sentence lays out the basic idea Einstein wants to develop in his theory of special relativity. See "On the

Electrodynamics of Moving Bodies," *Annalen der Physik* 19,
1905. In *CPAE*, Vol. 2, Doc. 23

$E = mc^2$.

Statement of the equivalence of mass and energy—energy
equals mass times the speed of light squared—which
opened up the atomic age, though Einstein had no premonition or foresight about it at the time. The original statement was: "If a body emits the energy L in the form of radiation, its mass decreases by L/V^2." (Originally in "Ist die
Trägheit eines Körpers von seinem Energieinhalt abhän-
gig?" *Annalen der Physik* 18 [1905], 639–641. See Stachel et al.,
Einstein's Miraculous Year, 161, for a translation of this paper.)
Note that Einstein used L (among other letters) to denote energy at least until 1912, when, in his "Manuscript on the
Special Theory of Relativity" (see *CPAE*, Vol. 4, Doc. 1), he
crossed out the L and substituted E in equations 28 and 28'
of the handwritten manuscript. (See the facsimile edition
of the manuscript published by George Braziller with the
Safra Foundation and the Israel Museum, Jerusalem [1996],
119 and 121; and *CPAE*, Vol. 4, Doc. 1, 58–59.) Interestingly,
as pointed out to me by Ralph Baierlein, he did the same
in a manuscript as late as January 1922 (see *CPAE*, Vol. 7,
Doc. 31, p. 259).

The equation derives from the special theory of
relativity, which played a decisive role in the investigation and development of nuclear energy. A mass
can be converted into a vast amount of energy
(i.e., when a particle is released from an atom it is
converted to energy), demonstrating a fundamental
relationship in nature. The theory also introduced a
new definition of space and time. For direct experimental evidence in its favor, however, it had to wait
twenty-five years, when the conversion of mass
into energy was confirmed in the study of nuclear

reactions; time dilation was not directly proved until 1938.

(*Note*: The following six quotations are out of chronological order, but I placed them here because they reflect Einstein's thoughts leading to the 1905 special theory.)

What if one were to run after a ray of light? . . . What if one were riding on the beam? . . . If one were to run fast enough, would it no longer move at all? . . . What is the "velocity of light"? If it is in relation to something, this value does not hold in relation to something else which is itself in motion.

Based on a conversation with psychologist Max Wertheimer in 1916, in which Einstein attempted to explain his thought process when he formulated the special theory of relativity. See Wertheimer, *Productive Thinking* (1945; reprinted by Harper, 1959), 218.

*For me it was always incomprehensible why the theory of relativity, whose concepts and problems are so far removed from practical life, should have found such a lively, even passionate resonance in the widest circles of the population for such a long time.

From the foreword to Philipp Frank, *Einstein*, ca. 1942. See also Rowe and Schulmann, *Einstein on Politics*, 130. Einstein Archives 28-581

After ten years of reflection, such a principle resulted from a paradox upon which I had already hit at the age of sixteen: If I pursue a beam of light with the velocity c (velocity of light in a vacuum), I should observe such a beam of light as a spatially oscillating electromagnetic field at rest. . . . From the very beginning it appeared to me intuitively clear that, judged from the standpoint of such an observer, everything would have to happen according to the same laws as for an observer who, relative to earth, was at rest.

Written in 1946 for "Autobiographical Notes," 53

*I despaired of the possibility of discovering the true laws by means of constructive efforts based on known facts. The longer and more despairingly I tried, the more I came to the conviction that only the discovery of a universal formal principle could lead us to assured results. The example I saw before me was thermodynamics.

Ibid.

The special theory of relativity owes its origin to Maxwell's equations of the electromagnetic field. Conversely, the latter can be grasped formally in satisfactory fashion only by way of the special theory of relativity.

Ibid., 63

That the special theory of relativity is only the first step of a necessary development became completely clear to me only in my efforts to represent gravitation in the framework of this theory.

Ibid.

Five or six weeks elapsed between the conception of the idea for the special theory of relativity and the completion of the relevant publication.

To Carl Seelig, March 11, 1952. Einstein Archives 39-013

My direct path to the special theory of relativity was mainly determined by the conviction that the electromotive force induced in a conductor moving in a magnetic field is nothing other than an electric field.

From a message read at a celebration of the centennial of Albert Michelson's birth, December 19, 1952, at Case Institute. See Stachel et al., *Einstein's Miraculous Year*, 111. Einstein Archives 1-168

According to the assumption considered here, in the propagation of a light ray emitted from a point source, the energy is not distributed continuously over ever-increasing volumes of space, but consists of a finite number of energy quanta localized at points of space that move without dividing and can be absorbed or generated only as complete units.

From "On a Heuristic Point of View Concerning the Production and Transformation of Light," March 1905. See Stachel et al., *Einstein's Miraculous Year*, 178. Considered by some to be the most revolutionary sentence written by a

twentieth-century physicist; see Fölsing, *Albert Einstein*,
143. *CPAE*, Vol. 2, Doc. 14

*We cannot ascribe *absolute* meaning to the concept
of simultaneity; instead, two events that are simultaneous when observed from some particular coordinate system can no longer be considered simultaneous when observed from a system that is moving
relative to that system.

From "On the Electrodynamics of Moving Bodies" (1905),
in Stachel et al., *Einstein's Miraculous Year*, 130. *CPAE*,
Vol. 2, Doc. 23

I've completely solved the problem. My solution
was to analyze the concept of time. Time cannot be
absolutely defined, and there is an inseparable relation between time and signal velocity.

Said to Michele Besso, May 1905, in reference to his forth-
coming publication, "On the Electrodynamics of Moving
Bodies," on the relativity principle in electrodynamics,
later to be called the special theory of relativity. Recalled
during Einstein's lecture in Kyoto, December 14, 1922. See
Physics Today (August 1982), 46

[I will send you] four papers. [The first] deals with
radiation and the energetic properties of light and is
very revolutionary. . . . The second paper determines
the true size of atoms by way of diffusion and the
viscosity of diluted solutions of neutral substances.
The third proves that, assuming the molecular theory of heat, bodies on the order of magnitude of

1/1000mm, when suspended in liquids, must already have an observable random motion that is produced by thermal motion. . . . The fourth paper is only a rough draft right now, and is about the electrodynamics of moving bodies that employs a modified theory of space and time.

To Conrad Habicht, May 1905, giving him a foretaste of Einstein's *annus mirabilis*, during which he published, at the age of twenty-six, altogether five important papers that ushered physics into a new era. See Stachel et al., *Einstein's Miraculous Year*, for a presentation and discussion of the papers. *CPAE*, Vol. 5, Doc. 27

*One more consequence of the electrodynamical paper has also occurred to me. The principle of relativity, together with Maxwell's equations, requires that mass be a direct measure of the energy contained in a body; light transfers mass. . . . The contemplation is amusing and attractive, but I don't know if the good Lord is laughing at it and leading me around by the nose.

To Conrad Habicht, Summer 1905. *CPAE*, Vol. 5, Doc. 28

From this we conclude that a balance-wheel clock located at the Earth's equator must go more slowly, by a very small amount, than a precisely similar clock situated at one of the poles under otherwise identical conditions.

From "On the Electrodynamics of Moving Bodies." Originally in "Zur Elektrodynamik bewegter Körper," *Annalen der Physik* 17 (1905), 891–921. See *CPAE*, Vol. 2, Doc. 23. This is the paper in which Einstein introduces special relativity. According to a letter I received from Professor Emeritus I. J. Good of Virginia Tech in Blacksburg, Einstein neglected to add that he was assuming the frame of reference of the observer at the pole. In other inertial frames of reference, the clock of the person on the equator seems to go more slowly than that of the person at the pole at least *some* of the time, but not necessarily *all* of the time, as Einstein seems to be saying. This lapse in exposition (or possibly a mistake) led physicist Herbert Dingle astray; he spent many years producing incorrect arguments against the special theory of relativity.

All our judgments in which time plays a role are judgments about simultaneous events. If I say, for example, "the train arrives here at 7," this means: the coincidence of the small hand of my watch with the number 7 and the arrival of the train are simultaneous events.

Ibid.

(1) The laws according to which the states of a physical system change do not depend on which of the two coordinate systems, in uniform relative motion, these laws refer to. (2) Every light ray moves in a "rest" coordinate system with a definite speed c, whether emitted from a stationary or moving force.

Ibid. According to Leopold Infeld (*Albert Einstein*, 24), these are the foundations on which relativity theory is based.

*So far we have applied the principle of relativity, i.e., the assumption that the physical laws are independent of the state of motion of the reference system, only to *nonaccelerated* reference systems. Is it conceivable that the principle of relativity also applies to systems that are accelerated relative to each other?

> First paragraph of Part 5, Sec. 17 of "The Principle of Relativity and the Conclusions Drawn from It" (1907), laying the groundwork for the general theory of relativity of 1915. *CPAE*, Vol. 2, Doc. 47

Thanks to my fortunate idea of introducing the relativity principle into physics, you (and others) now enormously overrate my scientific abilities, to the point where this makes me quite uncomfortable.

> To Arnold Sommerfeld, January 14, 1908. *CPAE*, Vol. 5, Doc. 73

A physical theory can be satisfactory only if its structures are composed of elementary foundations. The theory of relativity is ultimately as little satisfactory as, for example, classical thermodynamics was before Boltzmann had interpreted the entropy as probability.

> Ibid.

People who have been privileged to contribute something to the advancement of science should not let [arguments about priority] becloud their joy over the fruits of common endeavor.

> To Johannes Stark, February 22, 1908. A few days earlier, Einstein had expressed some annoyance that Stark failed to recognize Einstein's priority in regard to the relativistic relationship between mass and energy, which Stark had attributed to Max Planck in a paper in the *Physikalische Zeitschrift* in December 1907. See *CPAE*, Vol. 5, Doc. 88, and Doc. 70, n. 3

It seems that scientific distinction and personal qualities do not always go hand in hand. I value a harmonious person far more than the craftiest formula jockey or experimentalist.

> To Jakob Laub, March 16, 1910, lauding Laub's boss, Alfred Kleiner. *CPAE*, Vol. 5, Doc. 199

The more success the quantum theory has, the sillier it looks. How nonphysicists would scoff if they were able to follow the odd course of developments!

> To Heinrich Zangger, May 20, 1912, reflecting Einstein's early lack of faith in the quantum theory. *CPAE*, Vol. 5, Doc. 398

The "theory of relativity" is correct insofar as the two principles upon which it is based are correct. Since these do seem to be largely correct, the theory of relativity in its present form seems to represent an important advance. I do not think that it has hampered the further development of theoretical physics!

From "Reply to Comment by M. Abraham," August 1912.
CPAE, Vol. 4, Doc. 8

I am now working exclusively on the gravitation problem. . . . One thing is certain: never before in my life have I troubled myself over anything so much, and I have gained enormous respect for mathematics, whose more subtle parts I considered until now . . . as pure luxury! Compared with this problem, the original theory of relativity is child's play.

To Arnold Sommerfeld, October 29, 1912, indicating his difficulties with advanced mathematics in formulating the general theory of relativity, with which his friend Marcel Grossmann helped him. *CPAE*, Vol. 5, Doc. 421

I cannot find the time to write because I am occupied with truly great things. Day and night I rack my brain in an effort to penetrate more deeply into the things that I gradually discovered in the past two years and that represent an unprecedented advance in the fundamental problems of physics.

To Elsa Löwenthal, February 1914, about his work on an extension of his theory of gravitation, the first stage of which was published half a year earlier. *CPAE*, Vol. 5, Doc. 509

Nature is showing us only the tail of the lion, but I have no doubt that the lion belongs to it even though, because of its large size, it cannot totally reveal itself all at once. We can see it only the way a louse that is sitting on it would.

> To Heinrich Zangger, March 10, 1914, regarding his work on the general theory of relativity. *CPAE*, Vol. 5, Doc. 513

The principle of relativity can generally be phrased as: The laws of nature perceived by an observer are *independent* of his state of motion. . . . By combining the principle of relativity with the results of the constancy of light in a vacuum, one arrives by a purely deductive manner at what is called today "relativity theory." . . . Its significance lies in the fact that it provides conditions that every general law of nature must satisfy, for the theory teaches that natural phenomena are such that the laws do not depend on the state of motion of the observer to whom the phenomena are spatially and temporally related.

> *Vossische Zeitung*, April 26, 1914. *CPAE*, Vol. 6, Doc. 1

*A theorist goes astray in two ways:

1. The devil leads him by the nose with a false hypothesis. (For this he deserves our pity.)
2. His arguments are erroneous and sloppy. (For this he deserves a beating.)

> To H. A. Lorentz, February 3, 1915. *CPAE*, Vol. 8, Doc. 52

One should not pursue goals that are easily achieved. One must develop an instinct for what one can just barely achieve through one's greatest efforts.

To former student Walter Dällenbach, May 31, 1915, while giving him some advice on an electrical engineering project. *CPAE*, Vol. 8, Doc. 87

Professionally, scientists and mathematicians are strictly international-minded and guard carefully against any unfriendly measures taken against their colleagues living in hostile foreign countries. Historians and philologists, on the other hand, are mostly chauvinistic hotheads.

To H. A. Lorentz, August 2, 1915, on the atmosphere in Berlin, though Einstein spoke about a specific mind-set in Germany that was conditioned by historical circumstance. *CPAE*, Vol. 8, Doc. 103

In my personal experience I have hardly come to know the wretchedness of mankind better than as a result of this theory and everything connected to it. But it doesn't bother me.

To Heinrich Zangger, November 26, 1915, regarding the reception of the general theory of relativity. *CPAE*, Vol. 8, Doc. 152

The theory is beautiful beyond comparison. However, only *one* colleague has really been able to understand it and [use it].

Ibid. The colleague was David Hilbert.

Hardly anyone who truly understands it will be able to escape the charm of this theory.

From "Field Equations of Gravitation," November 1915, a paper further confirming the general theory of relativity by applying Riemann's curvature tensor. *CPAE*, Vol. 6, Doc. 25

Be sure you take a good look at them; they are the most valuable discovery of my life.

To Arnold Sommerfeld, December 19, 1915, regarding the equations in the above paper. *CPAE*, Vol. 8, Doc. 161

A concept exists for a physicist only when there is a possibility of finding out in a *concrete* case whether or not the concept applies.

This sentence appears in Einstein's popular account of relativity, *On the Special and the General Theory of Relativity* (1916; published in German in 1917) (see *CPAE*, Vol. 6, Doc. 42). Reprinted in English in *Relativity: The Special and the General Theory*. It is his comment on the assumption of the absolute nature of simultaneity (see *CPAE*, Vol. 9, Doc. 316, n. 3). Also quoted in a letter from Edouard Guillaume to Einstein, February 15, 1920 (*CPAE*, Vol. 9, Doc. 316).

*No fairer destiny could be allotted to any physical theory than that it should of itself point out the way to the introduction of a more comprehensive theory, in which it lives on as a limiting case.

From *Relativity: The Special and the General Theory*, 78

The theory of relativity is nothing but another step in the centuries-old evolution of our science, one which preserves the relationships discovered in the past, deepening their insights and adding new ones.

From "The Principal Ideas of the Theory of Relativity," written after December 1916. *CPAE*, Vol. 6, Doc. 44a, embedded in Vol. 7

The supreme task [*Aufgabe*] of the physicist is to arrive at those universal elementary laws from which the cosmos can be built up by pure deduction. There is no logical path to these laws; only intuition, resting on sympathetic understanding of experience, can reach them.

From "Motives for Research," a speech delivered at Max Planck's sixtieth birthday celebration, April 1918. Reprinted in *Ideas and Opinions*, 226, as "Principles of Research." See *CPAE*, Vol. 7, Doc. 7

The state of mind which enables a man to do work of this kind . . . is akin to that of the religious worshiper or the lover; the daily effort comes from no deliberate intention or program, but straight from the heart.

Ibid., 227

In regard to his subject matter . . . the physicist has to limit himself very severely: he must content himself with describing the most simple events that can be brought within the domain of our experience; all events of a more complex order are beyond the power of the human intellect to reconstruct with the subtle accuracy and logical perfection the theoretical physicist demands.

Ibid.

I believe with Schopenhauer that one of the strongest motives that leads men to art and science is escape from everyday life with its painful crudity and hopeless dreariness, from the fetters of one's own ever shifting desires. . . . A finely tempered nature longs to escape from personal life into the world of objective perception and thought.

Ibid.

The mainspring of scientific thought is not an external goal toward which one must strive, but the pleasure of thinking.

To Heinrich Zangger, ca. August 11, 1918. *CPAE*, Vol. 8, Doc. 597

For me, a hypothesis is a statement whose *truth* is temporarily assumed, but whose *meaning* must be beyond all doubt.

To Edward Study, September 25, 1918. *CPAE*, Vol. 8, Doc. 624

Nature rarely surrenders one of her magnificent secrets!

To Heinrich Zangger, June 1, 1919, regarding his further explorations into relativity theory. *CPAE*, Vol. 9, Doc. 52

*The quantum theory gives me a feeling very much like yours. One really ought to be ashamed of its success, because it has been obtained with the Jesuit maxim: "Let not thy left hand know what thy right hand doeth."

> To Max Born, June 4, 1919. In Born, *Born-Einstein Letters*, 10. *CPAE*, Vol. 9, Doc. 56

Lecturing on quantum theory is not for me. Though I have labored much with it, I have gained little insight into it.

> To Walter Dällenbach, ca. July 1, 1919. *CPAE*, Vol. 9, Doc. 66

Dear Mother, Today I have some happy news. H. A. Lorentz telegraphed me that the English expeditions [led by Arthur Eddington] have really verified the deflection of light by the sun.

> To Pauline Einstein, September 27, 1919. *CPAE*, Vol. 9, Doc. 113. Some writers have claimed that Arthur Eddington fudged his data in compiling the results of his experiments. During the eclipse expeditions to the island of Principe that set out to prove general relativity, he threw out two-thirds of the sixteen photographic plates that seemed to support Newton over Einstein. Some researchers think that the mathematical formula that Eddington used to reach the star-beam displacement was also biased. In time, of course, Eddington was vindicated when others obtained better results and proved Einstein correct, anyway. For a discussion of this subject, see Daniel Kennefick, "Testing Relativity from the 1919 Eclipse," *Physics Today*, March 2009, 37–42.

The most important consequence of the special theory of relativity concerned the inert masses of corporeal systems. It became evident that the inertia of a system necessarily depends on its energy content, and this led straight to the notion that inert mass is simply latent energy. The principle of the conservation of mass lost its independence and became joined with that of the conservation of energy.

> From "What Is the Theory of Relativity," written at the request of *The Times* (London), November 28, 1919. *CPAE*, Vol. 7, Doc. 25

When we say that we understand a group of natural phenomena, we mean that we have found a constructive theory that embraces them.

> Ibid.

I believe that we can promote research effectively in the area of the general theory of relativity even without use of special public funds, if the country's observatories and astronomers would simply place a portion of their equipment and labor at the service of this cause.

> To Konrad Haenisch, the German Minister of Education, December 6, 1919, after being informed that the National Treasury had reserved 150,000 marks to support research in general relativity. *CPAE*, Vol. 9, Doc. 194

I am convinced that the redshift of spectrum lines is an absolutely convincing consequence of relativity theory. If it were proved that this effect did not exist in nature, then the whole theory would have to be abandoned.

To Arthur Eddington, December 15, 1919. *CPAE*, Vol. 9, Doc. 216

[A researcher] adapts to the facts by intuitive selection of the possible theories based upon axioms.

From "Induction and Deduction in Physics," *Berliner Tage-blatt*, December 25, 1919. See also *CPAE*, Vol. 7, Doc. 28

The simplest picture one can form about the creation of an empirical science is along the lines of an inductive method. Individual facts are selected and grouped together such that their lawful connection becomes clearly apparent. By grouping these laws together, one can achieve other more general laws until a more or less uniform system for the available individual facts has been established. . . . However . . . the big advances in scientific knowledge originated this way only to a small degree. For, if a researcher were to approach things without a preconceived opinion, how would he be able to pick the facts from the tremendous richness of the most complicated experiences that are simple enough to reveal their connections through laws?

Ibid.

The truly great advances in our understanding of nature originated in a way almost diametrically opposed to induction. The intuitive grasp of the essentials of a large complex of facts leads the scientist to the postulation of a hypothetical basic law, or several such laws. From these laws, he derives his conclusions, . . . which can then be compared to experience. Basic laws (axioms) and conclusions together form what is called a "theory." Every expert knows that the greatest advances in natural science . . . originated in this manner, and that their basis has this hypothetical character.

Ibid.

The *truth* of a theory can never be proven, for one never knows if future experience will contradict its conclusions.

Ibid.

When two theories are available and both are compatible with the given arsenal of facts, then there are no other criteria to prefer one over the other except the intuition of the researcher. Therefore one can understand why intelligent scientists, cognizant both of theories and of facts, can still be passionate adherents of opposing theories.

Ibid.

Then I would have had to pity our dear God. The theory is correct all the same.

In answer to the question of doctoral student Ilse Rosenthal-Schneider, in 1919, about how he would have reacted if his general theory of relativity had not been confirmed experimentally that year by Arthur Eddington and Frank Dyson. As quoted by her in Rosenthal-Schneider, *Reality and Scientific Truth*, 74

*Why should we give preference to coordinate systems in uniform motion? Any motion should be permissible. What does Nature care about our reference systems?

In ibid., 91

[Constructive theories], from a relatively simple fundamental formalism, attempt to explain the more complex phenomena. . . . [Theories of principle, on the other hand,] are based on empirically discovered general properties of natural processes, on principles from which mathematically formulated criteria follow and that individual processes or their theoretical models must observe.

Einstein's formulation of two kinds of scientific theories, 1919; he regarded constructive theories as more important, though each had its advantages. "What Is the Theory of Relativity?" *CPAE*, Vol. 7, Doc. 26

Some time will probably still have to elapse before the [spectral] problem is completely resolved. But I have full confidence in the relativistic idea. Once all the sources of error have been eliminated (indirect light source), it is sure to come out right.

> To Paul Ehrenfest, April 7, 1920, on general relativity. *CPAE*, Vol. 9, Doc. 371

Concepts are simply empty when they stop being firmly linked to experiences. They resemble social climbers who are ashamed of their origins and want to deny them.

> To Hans Reichenbach, June 30, 1920. *CPAE*, Vol. 10, Doc. 66

This world is a remarkable house of fools. At present every coachman and every waiter is debating whether the relativity theory is correct. Their conviction about that is determined by the political party to which they belong.

> To Marcel Grossmann, September 12, 1920, expressing his surprise at the widespread interest in the theory of general relativity. Because his theory was not understood by most people, Einstein became an even more mysterious figure. He continued to refer to the public spectacle as the "relativity circus." *CPAE*, Vol. 10, Doc. 148

It is my inner conviction that the development of science seeks in the main to satisfy the longing for pure knowledge.

> 1920. Quoted by Moszkowski, *Conversations with Einstein*, 173

The word "discovery" in itself is regrettable. For discovery is equivalent to becoming aware of a thing which is already formed; this links up with proof, which no longer bears the character of "discovery" but, in the final analysis, of the means that leads to discovery. . . . Discovery is really not a creative act.

> Ibid., 95

The aspect of knowledge that has not yet been laid bare gives the investigator a feeling akin to that experienced by a child who seeks to grasp the masterly way in which adults manipulate things.

> Ibid., 46

*I'm now pretty much fed up with relativity! Even such a thing wears thin when one becomes too preoccupied with it.

> To Elsa Einstein, January 8, 1921. *CPAE*, Vol. 12, Doc. 12

As far as the laws of mathematics refer to reality, they are not certain; and as far as they are certain, they do not refer to reality.

> From "Geometry and Experience," an address to the Prussian Academy of Sciences, Berlin, January 27, 1921. In Einstein, *Sidelights on Relativity* (1922; reprint, New York: Dover, 1983), 28. (In the 1996 edition of this book, in place of "the laws of mathematics," I had earlier used Philipp Frank's word "geometry," misquoted in *Einstein: His Life and Times*, 177. Thanks to a Czech reader for the correction.)

One reason why mathematics enjoys special esteem, above all other sciences, is that its laws are absolutely certain and indisputable, while those of all other sciences are to some extent debatable and in constant danger of being overthrown by newly discovered facts.

Ibid., 27

We may in fact regard [geometry] as the most ancient branch of physics. . . . Without it I would have been unable to formulate the theory of relativity.

Ibid., 32–33

*The Lord does it the way *he* wants to and will not be dictated to.

To Arnold Sommerfeld, March 9, 1921, in presenting a supplementary general relativity equation to him, but being uncertain of its physical value. *CPAE*, Vol. 12, Doc. 89

*No man of culture or knowledge has any animosity toward my theories. Even the physicists opposed to the theory are animated by political motives.

Quoted in the *New York Times*, April 3, 1921, 1, 13. See also Illy, *Albert Meets America*, 30

*Without the discoveries of every one of the great men of physics, those who laid down preceding laws, relativity would have been impossible to conceive, and there would have been no basis for it. . . . The four men who laid the foundations of physics on which I have been able to construct my theory are Galileo, Newton, Maxwell, and Lorentz.

Quoted in the *New York Times*, April 4, 1921, 5. See also Illy, *Albert Meets America*, 41–42

*The practical man need not worry about [relativity theory]. From the philosophical aspect, however, it has importance, as it alters the conceptions of time and space which are necessary to philosophical speculations and conceptions.

Ibid.

*A clock on the periphery moves slower than a clock at the center for an observer on the blackboard. It must go more slowly, it has been shown, for the observer on the disk, hence the conclusion that as soon as a gravitational field is at hand, of which this is a special case, then clocks at different places run at different speeds.

From a lecture at the City College of New York, April 20, 1921. *New York Times*, April 21, 1921, 12. See also Illy, *Albert Meets America*, 108

*Everywhere I go someone asks me that question. It is absurd. Anyone who has had sufficient training in science can readily understand the theory. There is nothing amazing or mysterious about it. It is so simple to minds trained along that line and there are many such in the United States.

> From an interview with the *Chicago Daily Tribune*, published May 3, 1921, sec. 1, 3, in response to the question if it is true that only twelve men can understand his theories of relativity. See also Illy, *Albert Meets America*, 147

The Lord God is subtle, but malicious he is not.

> Originally said in German to Princeton University mathematics professor Oswald Veblen, May 1921, while Einstein was in Princeton for a series of lectures, upon hearing that an experimental result by Dayton C. Miller of Cleveland, if true, would contradict his theory of gravitation; but the result turned out to be false. Some say by this remark Einstein meant that Nature hides her secrets by being subtle, while others say he meant that Nature is mischievous but not bent on trickery. The above translation was popularized by Abraham Pais, but the original German word "raffiniert" is difficult to translate. Other appropriate adjectives, while not as pretty as "subtle," are crafty, wily, tricky, cunning.
>
> Permanently inscribed in stone above the fireplace in the faculty lounge, 202 Jones Hall (called Fine Hall until Princeton's new mathematics building with the same name was constructed) as "Raffiniert ist der Herr Gott, aber boshaft ist Er nicht" ("Herr Gott" should be "Herrgott"). Quoted widely in various translated versions, for example, in Pais, *Subtle Is the Lord*; Frank, *Einstein: His Life and Times*, 285; and Hoffmann, *Albert Einstein: Creator and Rebel*, 146

I have second thoughts. Maybe God *is* malicious.

> To Valentine Bargmann and Peter Bergmann, later in Princeton, meaning that God makes us believe we have

understood something that in reality we are far from understanding. See Bargmann, "Working with Einstein," in *Some Strangeness in the Proportion,* ed. Harry Woolf (Addison-Wesley, 1980), 480–481.

Now to the term "relativity theory." I admit that it is unfortunate, and has given occasion to philosophical misunderstandings.

To E. Zschimmer, September 30, 1921, referring to Max Planck's term for his theory, which stuck despite his unhappiness with it. He would have preferred "theory of invariants," which he felt better described the *method,* if not the content. See Holton, *The Advancement of Science,* 69, 110, 312, n. 21. Einstein Archives 24-156

*I build on Newton, I do not annul him. So, only through misinterpretation can my theory have a poetic significance, leading, without my intention or permission, to misunderstandings that are construed around it and to which it is forced to conform. Let us no longer talk of the political concerns or principles that some would like to connect to it.

From an interview with Aldo Sorani, *Il Messagero,* October 26, 1921. *CPAE,* Vol. 12, Appendix G

*The insights and methods developed by science serve practical purposes only indirectly and often only for future generations; but if we neglect science we will later lack the scientific workers who, by their broad vision and judgment, are able to

create new niches in the economy or adapt to new challenges.

From "The Plight of German Science," in *Neue Freie Presse* (Vienna), December 25, 1921. *CPAE*, Vol. 7, Doc. 70

Relativity is a purely scientific matter and has nothing to do with religion.

In response to the question of Randall Thomas Davidson, the Archbishop of Canterbury, about "what effect relativity would have on religion," London, 1921. Quoted in Frank, *Einstein: His Life and Times*, 190

*The working theoretical physicist is not to be envied, because Mother Nature, or more precisely an experiment, is a resolute and seldom friendly referee of his work. She never says "yes" to a theory, but only "maybe" under the best of circumstances, and in most cases simply "no." If an experiment verifies a theory, it is still a "maybe"; if it doesn't, it is a "no."

Perhaps a play on the universal notion that when a woman says "no" she means "maybe." In "Theoretische Bemerkungen zur Supraleitung der Metalle," in *Het natuurkundig laboratorium der Rijksuniversiteit te Leiden in de jaren 1904–1922*, November 11, 1922 (Leiden: Ijdo, 1922), 429. (Thanks to József Illy for sending me this gem.)

There is always a certain charm in tracing the evolution of theories in the original papers; often such study offers deeper insights into the subject matter than the systematic presentation of the final result, polished by the words of many contemporaries.

From the foreword to the Japanese edition of Einstein's papers, written in German, dated December 12, 1922, published May 1923

I was sitting in the patent office in Bern when all of a sudden a thought occurred to me: if a person falls freely, he won't feel his own weight. I was startled. This simple thought made a deep impression on me. It impelled me toward a theory of gravitation.

From his Kyoto lecture, December 14, 1922. Translated into English by Y. A. Ono in *Physics Today*, August 1932, from notes taken by Yon Ishiwara.

Describing the physical laws without reference to geometry is similar to describing our thoughts without words.

Ibid.

The theory of relativity states: The laws of nature are to be formulated free of any specific coordinates because a coordinate system does not conform to anything real. The simplicity of a hypothetical law is to be judged only according to its generally covariant form. . . . The laws of nature have never had and still do not have a preferential coordinate system. . . . The theory of relativity claims only that the *general* laws of nature are the same with respect to any system.

From an article in *Annalen der Physik* 69 (1922), 438. Einstein Archives 1-016

*Give the boy his Ph.D. He can't do much damage with a Ph.D. in physics!

To Paul Langevin, ca. 1922, in urging him to accept Prince Louis de Broglie's doctoral dissertation claiming that matter can be considered to have a dual particle/wave nature, as had been established for light. Quoted in Bulent Atalay, *Math and the Mona Lisa* (Washington, D.C.: Smithsonian Books, 2004). Thanks to Tom Gilb for sending this gem to me, and to Atalay for sending me the book. The story may be apocryphal even though it is widely circulated by respected physicists, since I could not find it in any of Einstein's letters to Langevin. The closest was a letter of December 16, 1924 (Einstein Archives 15-377), where he writes that de Broglie has "lifted a corner of the great veil," that the theory fitted in with something he was working on, and that he would discuss the subject with others. De Broglie introduced his theory in notes in 1923, defended his dissertation on November 25, 1924 (three weeks before Einstein wrote this particular letter), and published it in 1925 in *Ann. de Phys.*, ser. 10, vol. 3. As it turned out, the theory was mathematically equivalent to the Heisenberg theory, which, according to the American Institute of Physics Web site, Einstein distrusted.

In seeking an integrated theory, the intellect cannot rest contentedly with the assumption that there are two distinct fields, totally independent of each other by their nature.

From his delayed Nobel Lecture, written June 11, 1923, and delivered July 1923 in Göteborg. This statement foretold Einstein's lifelong search for a unified field theory of gravity and electromagnetism. See *Les Prix Nobel en 1921–1922* (Stockholm, 1923). Einstein Archives 1-027

After a certain high level of technical skill is achieved, science and art tend to coalesce in esthetics, plasticity, and form. The greatest scientists are artists as well.

Remark made in 1923. Recalled by Archibald Henderson, *Durham Morning Herald*, August 21, 1955. Einstein Archives 33-257

The more one chases after quanta, the better they hide themselves.

To Paul Ehrenfest, July 12, 1924, expressing his frustration over quantum theory. Einstein Archives 10-089

My interest in science was always essentially limited to the study of principles. . . . That I have published so little is due to this same circumstance, as the great need to grasp principles has caused me to spend most of my time on fruitless pursuits.

To Maurice Solovine, October 30, 1924. Published in *Letters to Solovine*, 63. Einstein Archives 21-195

There are those with a good nose for fundamental physical insights [*Prinzipienfuchser*] and there are those who have great technical ability [*Virtuosen*]. . . .

All three of us [Einstein, Bohr, Ehrenfest] belong to the first kind and (at least the two of us) have little technical talent. Thus the effect when encountering outstanding virtuosos (Born or Debye): discouragement. But it is similar the other way around.

> To Paul Ehrenfest, September 18, 1925. Quoted in the German edition of Fölsing, *Albert Einstein*, 552. Einstein Archives 10-111

Quantum mechanics is certainly imposing. But an inner voice tells me that this is not yet the real thing. The theory yields much, but it hardly brings us closer to the Old One's secrets. I, in any case, am convinced that He does not play dice.

> To Max Born, December 4, 1926. In Born, *Born-Einstein Letters*, 88. Einstein Archives 8-180. The popular version of the last sentence is "God does not play dice with the universe."

It is only in quantum theory that Newton's differential method becomes inadequate, and indeed strict causality fails us. But the last word has not yet been said.

> Letter to the Royal Society (U.K.) on the occasion of the Newton bicentenary, March 1927. Reprinted in *Nature* 119 (1927), 467. Einstein Archives 1-060

All physical theories, their mathematical expressions notwithstanding, ought to lend themselves to so simple a description that even a child could understand them.

From a conversation in 1927; recalled by Louis de Broglie
in *Nouvelles perspectives en microphysique*, Paris, 1956.
(Trans. New York: Basic Books, 1962, 184.) Also in Clark,
Einstein, 344

*The best thing, which I have been pondering and
figuring out for days on end and half the night, is
now complete before me, condensed into seven
pages and titled "A Unified Field Theory."

To Michele Besso, January 5, 1929. Translated in Neffe, *Einstein*, 351. Einstein Archives 7-102

I admire to the highest degree the achievement
of the younger generation of physicists which goes
by the term quantum mechanics, and believe in
the deep level of truth of that theory; but I believe
that its limitation to *statistical laws* will be a temporary one.

From a speech on June 28, 1929, on acceptance of the
Planck Medal. Quoted in *Forschungen und Fortschritte* 5
(1929), 248–249

The main source of all technological achievements is
the divine curiosity and playful drive of the tinker-
ing and thoughtful researcher, as much as it is the
creative imagination of the inventor.

From a radio broadcast in Berlin opening the German
Radio Exhibition, August 22, 1930. Transcribed by Friedrich Herneck in *Die Naturwissenschaften* 48 (1961), 33. Einstein Archives 4-044

Those who thoughtlessly make use of the miracles of science and technology, without understanding more about them than a cow eating plants understands about botany, should be ashamed of themselves.

Ibid.

*Until our era, people of different nations got to know one another almost exclusively by way of the distorting mirror of their own daily press. Radio shows people to one another in a vibrant way . . . and thereby contributes to eradicating the feeling of mutual alienation that can easily turn to mistrust and hostility.

Ibid.

The scientist finds his reward in what Henri Poincaré calls the joy of comprehension, and not in the possibilities of application to which any discovery may lead.

From "A Socratic Dialogue," an exchange of opinions among Einstein, James Murphy, and J.W.N. Sullivan, probably in 1930. As quoted in the epilogue to Planck, *Where Is Science Going?* (New York, 1932), 211. Parts of the dialogue are also in *Forum and Century* 83 (June 1930), 373–379, under the title "Science and God: A Dialogue."

A dictatorship means muzzles all round, and consequently stultification. Science can flourish only in an atmosphere of free speech.

From "Science and Dictatorship," in *Dictatorship on Its Trial*, ed. Otto Forst de Battaglia, trans. Huntley Paterson (London: George G. Harrop, 1930), 107. This contributed essay consists of only these two sentences. Einstein Archives 46-218

Concern for man himself and his fate must always form the chief objective of all technological endeavors . . . in order that the creations of our minds shall be a blessing and not a curse to mankind. Never forget this in the midst of your diagrams and equations.

From an address entitled "Science and Happiness," presented at the California Institute of Technology, Pasadena, February 16, 1931. Quoted in the *New York Times*, February 17 and 22, 1931. Einstein Archives 36-320

Why does this magnificent applied science, which saves work and makes life easier, bring us so little happiness? The simple answer: because we have not yet learned to make a sensible use of it.

In reference to technology. Ibid.

*The belief in an external world independent of the perceiving subject is the basis of all natural science. Since, however, sense perception only gives information of this external world or of "physical reality" indirectly, we can only grasp the latter by speculative means. It follows from this that our notions of physical reality can never be final.

From "Maxwell's Influence on the Evolution of the Idea of Physical Reality," in *James Clerk Maxwell: A Commemorative Volume* (Cambridge, U.K.: Cambridge University Press, 1931). See also *Ideas and Opinions*, 266

*Thus the partial differential equation entered theoretical physics as a handmaid but has gradually become a mistress.

Ibid. *Ideas and Opinions*, 268

I believe that the present fashion of applying the axioms of physical science to human life is not only entirely a mistake but has also something reprehensible about it.

On a "worldview" of relativity and the gross abuse of physical science in areas in which it is not applicable. Ibid.; also quoted by Loren Graham in Holton and Elkana, *Albert Einstein: Historical and Cultural Perspectives*, 107

*Science as something already in existence, already completed, is the most objective, impersonal thing that we humans know. Science as something coming into being, as a goal, is just as subjectively, psychologically conditioned as are all other human endeavors.

From an address to students at UCLA, February 1932. In *Builders of the Universe* (Los Angeles: U.S. Library Association, 1932), 91

*It can scarcely be denied that the supreme goal of all theory is to make the irreducible basic elements as simple and as few as possible without having to surrender the adequate representation of a single datum of experience.

> From "On the Method of Theoretical Physics," the Herbert Spencer Lecture, Oxford, June 10, 1933. This is the Oxford University Press version. The words "simple," "simplest," and "simplicity" recur throughout the lecture. The version reprinted in 1954 in *Ideas and Opinions*, 272, is a bit different. This sentence may be the origin of the much-quoted sentence that "everything should be as simple as possible, but not simpler," and its variants. The latter version, which appears in the July 1977 issue of *Reader's Digest*, of course can't be taken literally. The October 1938 *Reader's Digest* article with the word "simplicity" in its title uses the word only to describe Einstein as a human being. This article, incidentally, is full of biographical errors.

*Our experience hitherto justifies us in believing that nature is the realization of the simplest conceivable mathematical ideas. I am convinced that we can discover by means of purely mathematical constructions the concepts and the laws connecting them with each other, which furnish the key to the understanding of natural phenomena.

> Ibid. *Ideas and Opinions*, 274

The creative principle [of science] resides in mathematics.

> Ibid.

The years of anxious searching in the dark for a truth that one feels but cannot express, the intense desire and the alternations of confidence and misgiving until one achieves clarity and understanding, can be understood only by those who have experienced them.

From a lecture at the University of Glasgow, June 20, 1933. Published in *The Origins of the Theory of Relativity*; reprinted in *Mein Weltbild*, 138; and in *Ideas and Opinions*, 289–290

It is not the *result* of scientific research that ennobles humans and enriches their nature, but the *struggle to understand* while performing creative and open-minded intellectual work.

From "Good and Evil," 1933. Published in *Mein Weltbild* (1934), 14; reprinted in *Ideas and Opinions*, 12

*He [a mathematician] has shown little psychological insight. Mathematicians are often so. They think logically, but they lack an organic connection.

To Stephen Wise, June 9, 1934. Einstein Archives 35-150

*The idea that there exist two structures of space independent of each other, the metric-gravitational field and the electromagnetic, [is] intolerable to the theoretical spirit. We are prompted to the belief that both sorts of fields must correspond to a unified structure of space.

From "The Problem of Space, Ether, and the Field in Physics," in *Essays in Science* (1934), 74. Also in *Ideas and Opinions*, 285

[The likelihood of transforming matter into energy] is something like shooting birds in the dark in a country where there are only a few birds.

> Remark at a January 1935 press conference, three years before the atom was successfully split to cause fission. Quoted in *Literary Digest,* January 12, 1935. Also quoted in Nathan and Norden, *Einstein on Peace,* 290, which warns that the account should be regarded with caution.

I myself have only little motivation to write general things because I feel a strong alienation from the generation with which I will share the rest of my days. I would rather bury myself in contemplating basic scientific problems, particularly those that in my opinion currently strongly deviate from prevailing work. I don't think one will establish physics successfully using fundamental statistical foundations.

> To Bertrand Russell, January 27, 1935. Einstein Archives 33-161

The general public may be able to follow the details of scientific research to only a modest degree; but it can register at least one great and important notion: the confidence that human thought is dependable and natural law is universal.

> From "Science and Society," 1935. Reprinted in *Einstein on Humanism,* 13. Einstein Archives 28-342

Scientific research is based on the assumption that all events, including the actions of mankind, are determined by the laws of nature.

> To Phyllis Wright, January 24, 1936. Einstein Archives 52-337

All of science is nothing more than the refinement of everyday thinking.

> From "Physics and Reality," *Journal of the Franklin Institute* 221, no. 3 (March 1936), 349–382. Reprinted in *Ideas and Opinions*, 290

The aim of science is, on the one hand, a comprehension, as *complete* as possible, of the connection between the sense experiences in their totality, and, on the other hand, the accomplishment of this aim by the use of a *minimum of primary concepts and relations.*

> Ibid., 293

It is always a blessing when a great and beautiful conception is proven to be in harmony with reality.

> To Sigmund Freud, April 21, 1936, on Freud's ideas. Einstein Archives 32-566

We (Mr. Rosen and I) sent our publication to you without the authorization that you may show it to other specialists before it is printed. I do not see any reason to follow your anonymous reviewer's recommendations (which incidentally are erroneous). In view of the foregoing, I will consider having the work published elsewhere.

To the editor of the *Physical Review*, July 27, 1936. The article, "On Gravitational Waves," with Nathan Rosen, was later published in the *Journal of the Franklin Institute* 223 (1937), 43–54. Einstein Archives 19-087

I still struggle with the same problems as ten years ago. I succeed in small matters but the real goal remains unattainable, even though it sometimes seems palpably close. It is hard yet rewarding: hard because the goal is beyond my abilities, but rewarding because it makes one oblivious to the distractions of everyday life.

To Otto Juliusburger, September 28, 1937. Einstein Archives 38-163

*I'm still working passionately, though most of my intellectual off-spring are ending up prematurely in the cemetery of disappointed hopes.

To Heinrich Zangger, February 27, 1938. Einstein Archives 40-105

Physical concepts are free creations of the human mind and are not, however it may seem, uniquely determined by the external world.

From *The Evolution of Physics*, with Leopold Infeld (1938)

*According to the theory of relativity, there is no essential distinction between mass and energy. Energy has mass and mass represents energy. Instead of two conservation laws, we have only one, that of mass-energy.

Ibid., 208

*Without the belief that it is possible to grasp reality with our theoretical constructions, without the belief in the inner harmony of our world, there could be no science. This belief is and always will remain the fundamental motive for all scientific creation.

Ibid., 313

*The results gained thus far concerning the splitting of the atom do not justify the assumption that the atomic energy released in the process could be economically utilized. Yet, there can hardly be a physicist with so little intellectual curiosity that his interest in this important subject could become impaired because of the unfavorable conclusion to be drawn from past experimentation.

Statement for the *New York Times*, March 14, 1939. See Schweber, *Einstein and Oppenheimer*, 45

*Science is the attempt to make the chaotic diversity of our sense-experience correspond to a logically uniform system of thought. In this system, single experiences must be correlated with the theoretic structure in such a way that the resulting coordination is unique and convincing. . . . The sense experiences are the given subject matter, but the theory that shall interpret them is man-made. It is . . . hypothetical, never completely final, always subject to question and doubt.

From "The Fundamentals of Theoretical Physics," *Science* 91 (May 24, 1940), 487–492. Reprinted in *Ideas and Opinions*, 323–335

What we call physics comprises that group of natural sciences which base their concepts on measurements, and whose concepts and propositions lend themselves to mathematical formulations.

Ibid.

There has always been an attempt to find a unifying theoretical basis for all these [various branches of physics] . . . from which all the concepts and relationships among the individual disciplines might be derived by a logical process. This is what we mean by the search for a foundation of the whole of physics. The confident belief that this ultimate goal may be reached is the wellspring of the passionate devotion that has always motivated the researcher.

Ibid.

You cannot love a car the way you love a horse. The horse, unlike a machine, compels human emotions.

A machine disregards human feelings. . . . Machines make our life impersonal, stunt certain qualities in us, and create an impersonal environment.

> From a conversation recorded by Algernon Black, Fall 1940. Einstein forbade the publication of this conversation. Einstein Archives 54-834

Although it is true that it is the goal of science to discover rules which permit the association and foretelling of facts, this is not its only aim. It also seeks to reduce the connections discovered to the smallest possible number of mutually independent conceptual elements. It is in this striving after the rational unification of the manifold that it encounters its greatest successes.

> From "Science, Philosophy and Religion," a symposium published by the Conference on Science, Philosophy and Religion in Their Relation to the Democratic Way of Life, New York, 1941. Reprinted in *Ideas and Opinions*, 48–49

*The supernational character of scientific concepts and scientific language is due to the fact that they have been set up by the best brains of all countries and all times. . . . They created the spiritual tools for the technical revolution which has transformed the life of mankind in the last century.

> From "The Common Language of Science," *Advancement of Science* 2, no. 5 (1941), 109–110. Reprinted in *Ideas and Opinions*, 336–337

It is hard to sneak a look at God's cards. But that he would choose to play dice with the world . . . is something I cannot believe for a single moment.

> To Cornel Lanczos, March 21, 1942, expressing his reaction to quantum theory, which refutes relativity theory by stating that an observer *can* influence reality, that events *do* happen randomly. Quoted in Hoffmann, *Albert Einstein: Creator and Rebel*, chapter 10; Frank, *Einstein: His Life and Times*, 208, 285; and Pais, *Einstein Lived Here*, 114. Einstein Archives 15-294. My favorite variant of this quotation, sent to me by a rabbi, is: "God doesn't play craps with the universe." Physicist Niels Bohr is said to have told Einstein, "Stop telling God what to do!"

I never understood why the theory of relativity, with its concepts and problems so far removed from practical life, should have met with such a lively, indeed passionate, reception among a broad segment of the public.

> Written in October 1942. Published in foreword to Frank, *Einstein: Sein Leben und seine Zeit*, 1979 German edition

*We have become Antipodean in our scientific expectations. You believe in the God who plays dice, and I in complete law and order in a world which objectively exists. . . . I firmly *believe*, but I hope that someone will discover a more realistic way, or rather a more tangible basis than has been my lot to find. Even the great initial success of the quantum theory does not make me believe in the fundamental dice-game, although I am well aware that our younger colleagues interpret this to be a consequence of senility.

> To Max Born, September 7, 1944. In Born, *Born-Einstein Letters*, 146. Einstein Archives 8-207

*A knowledge of the historic and philosophical background [of science] gives ... independence from prejudices ... from which most scientists are suffering. This independence created by philosophical insight is—in my opinion—the mark of distinction between a mere artisan or specialist and a real seeker after truth.

> To Robert Thornton, December 7, 1944. Einstein Archives 56-283

The entire history of physics since Galileo bears witness to the importance of the function of the theoretical physicist, from whom the basic theoretical ideas originate. A priori construction in physics is as essential as empirical facts.

> From a memo written with Hermann Weyl to the faculty of the Institute for Advanced Study, early 1945, recommending theoretician Wolfgang Pauli over Robert Oppenheimer for a professorship at the Institute. Pauli declined the offer, and Oppenheimer, who was offered the directorship in 1946, accepted. Quoted in Regis, *Who Got Einstein's Office?*, 135

The theory of relativity, as I developed it originally, still does not explain atomism and the quantum phenomena. And neither does it include a common mathematical formulation covering the phenomena of both the electromagnetic and gravitational fields. This demonstrates that the original formulation of the theory of relativity is not definitive . . . its means of expression are in process of evolution. . . . The task to which I am now giving my greatest efforts is to resolve the dualism between the theories of gravitation and electromagnetism, and to reduce them to the one and same mathematical form.

From an interview with Alfred Stern, *Contemporary Jewish Record* 8 (June 1945), 245–249

I am not a positivist. Positivism states that what cannot be observed does not exist. This conception is scientifically indefensible, for it is impossible to make valid affirmations of what people "can" or "cannot" observe. One would have to say "only what we observe exists," which is obviously false.

Ibid.

I sold myself body and soul to Science—a flight from the "I" and "we" to the "it."

To Hermann Broch, September 2, 1945. Quoted in Hoffmann, *Albert Einstein: Creator and Rebel*, 254. Einstein Archives 34-048.1

A scientific person will never understand why he should believe opinions only because they are written in a certain book. [Furthermore], he will never believe that the results of his own attempts are final.

> To J. Lee, September 10, 1945. Einstein Archives 57-061

*A theory which in its fundamental equations explicitly contains a non-basic constant would have to be somehow constructed from bits and pieces which are logically independent of each other; but I am confident that this world is not such that so ugly a construction is needed for its theoretical comprehension.

> To Ilse Rosenthal-Schneider, October 13, 1945. From a long exchange with his former Berlin student about the universal constants of nature and how they relate to reality. See Rosenthal-Schneider, *Reality and Scientific Truth*, 32–38. Einstein Archives 20-278. In a letter of April 23, 1949, he tells her that his remarks were not categorical assertions but conjectures grounded on intuition (ibid., p. 40).

One can organize to apply a discovery already made, but not to make one. Only a free individual can make a discovery. . . . Can you imagine an organization of scientists making the discoveries of Darwin?

> From an interview with Raymond Swing, "Einstein on the Atomic Bomb," part 1, *Atlantic Monthly*, November 1945

As a scientist, I believe that nature is a perfect structure, seen from the standpoint of reason and logical analysis.

To Raymond Benenson, January 31, 1946. Einstein Archives 56-505

I believe that the abominable deterioration of ethical standards stems primarily from the mechanization and depersonalization of our lives—a disastrous by-product of science and technology. Nostra culpa!

To Otto Juliusburger, April 11, 1946. Einstein Archives 38-228

In the beginning (if there was such a thing), God created Newton's laws of motion together with the necessary masses and forces. This is all; every-thing beyond this follows from the development of appropriate mathematical methods by means of deduction.

Written in 1946 for "Autobiographical Notes," 19

A theory is the more impressive the greater the sim-plicity of its premises, the more different kinds of things it relates, and the more extended its area of applicability.

Ibid., 33. Einstein often refers to the value of simple hy-potheses, believing they may become basic traits of future theoretical representations, as in the case of the emission and absorption of radiation. See *CPAE*, Vol. 6, Doc. 34; see also what may be a bad paraphrase of this general idea, the quotation on simplicity in "Attributed to Einstein" at the back of the book, and the note on the June 10, 1933, quota-tion above.

[Classical thermodynamics] is the only physical theory of universal content that I am convinced, within the framework of its basic concepts, will never be overthrown.

Ibid.

The pair Faraday-Maxwell has a most remarkable inner similarity with the pair Galileo-Newton—the former of each pair grasping the relations intuitively, and the second one formulating those relations exactly and applying them quantitatively.

Ibid., 35

*Even scholars of audacious spirit and fine instinct can be obstructed in the interpretation of facts by philosophical prejudices. The prejudice . . . consists in the faith that facts by themselves can and should yield scientific knowledge without free conceptual construction.

Ibid., 49

*Nature is constituted so that it is possible logically to lay down such strongly determined laws that within these laws only rationally, completely determined constants occur (not constants, therefore, whose numerical value could be changed without destroying the theory).

Ibid., 63

Physics is an attempt conceptually to grasp reality as something that is considered to be independent of its being observed. In this sense one speaks of "physical reality."

Ibid., 81

Equations of such complexity as are the equations of the gravitational field can be found only through the discovery of a logically simple mathematical condition that determines the equations completely or at least almost completely.

Ibid., 89

Science will stagnate if it is made to serve practical goals.

In answer to a question posed by the Overseas News Agency, January 20, 1947. Quoted in Nathan and Norden, *Einstein on Peace*, 402. Einstein Archives 28-733

*In truth, I never believed that the foundations of physics could consist of laws of a statistical nature.

Unpublished draft comment on Max Born's essay "Einstein's Statistical Theories" for Schilpp, *Albert Einstein: Philosopher-Scientist*, March 1947. Quoted in Stachel, *Einstein from B to Z*, 390. Einstein Archives 2-027

If God had been satisfied with inertial systems, he would not have created gravitation.

Said to Abraham Pais, 1947. See Pais, *A Tale of Two Continents*, 227

I believe that this is the God-given generalization of general relativity theory. Unfortunately, the Devil comes into play, since one cannot solve the [new] equations.

> On his most recent efforts to generalize general relativity to a so-called unified field theory. Ibid.

I do not like it when it can be done this way or that way. It should be: This way or not at all.

> On theories in general. Ibid.

*Anything we say about the real world must be, by necessity, hypothetical and a construction of the human mind. For what is immediately given to us are only sense perceptions. . . . As always, the conception of the existence of the real world is fundamental in physics. Without it there would be no borderline between psychology and physics. . . . Modern developments have changed nothing in this respect.

> To David Holland, June 25, 1948. Einstein Archives 9-305

*It is of great importance that the general public be given an opportunity to experience—consciously and intelligently—the efforts and results of scientific research. It is not sufficient that each result be taken up, elaborated, and applied by a few specialists in the field. Restricting the body of knowledge to a small group deadens the philosophical spirit of a people and leads to spiritual poverty.

From the foreword of September 10, 1948, to Lincoln Barnett's *The Universe and Dr. Einstein* (2d rev. ed. New York:
Bantam, 1957), 9

*Mathematics is a useful tool for social science. In
the actual solution of social problems, however,
goals and intentions are the dominant factors.

From an interview with Milton James for the *Cheyney Record,* the student newspaper of Cheyney State Teachers
College in Pennsylvania, October 7, 1948, in answer to the
question asking if mathematics can be a useful tool in solving social problems. Einstein Archives 58-013 to 58-015

In my scientific work, I am still hampered by the
same mathematical difficulties that have been making it impossible for me to confirm or refute my general relativistic field theory. . . . I won't ever solve it;
it will be forgotten and must later be rediscovered
again.

To Maurice Solovine, November 25, 1948. Published in *Letters to Solovine*, 105, 107. Einstein Archives 21-256, 80-865

*I believe the method of quantum mechanics is not a
satisfactory one in principle. Yet . . . I in no way want
to deny that this theory represents a significant, in a
sense definitive progress of physical knowledge. . . .
The foundation will be consolidated or replaced by
a more comprehensive one.

From "Quantum Mechanics and Reality," *Dialectica*, 1948.
Einstein Archives 1-151

I have little patience for scientists who take a board of wood, look for its thinnest part, and drill a great number of holes when the drilling is easy.

> Recalled by Philipp Frank in "Einstein's Philosophy of Science," *Reviews of Modern Physics* 21, no. 3 (July 1949): 349–355

The grand aim of all science is to cover the greatest number of empirical facts by logical deduction from the smallest number of hypotheses or axioms.

> Quoted in Lincoln Barnett, "The Meaning of Einstein's New Theory," *Life* magazine, January 9, 1950

*Time and again the passion for understanding has led to the illusion that man is able to comprehend the objective world rationally by pure thought without any empirical foundations—in short, by metaphysics. I believe that every true theorist is a kind of tamed metaphysicist, no matter how pure a "positivist" he may fancy himself to be.

> From "On the Generalized Theory of Gravitation," *Scientific American* 182, no. 4 (April 1950). See *Ideas and Opinions*, 342. Einstein Archives 1-155

*According to general relativity, the concept of space detached from any physical content does not exist. The physical reality of space is represented by a field whose components are continuous functions of four independent variables—the coordinates of space and time.

Ibid., 348

*It is the very essence of our striving for understanding that, on the one hand, it attempts to encompass the great and complex variety of man's experience and, on the other, it looks for simplicity and economy in the basic assumptions. The belief that these two objectives can go side by side is, in view of the primitive state of our scientific knowledge, a matter of faith. Without such faith, I could not have a strong and unshakable conviction about the independent value of knowledge.

From "Message to the Italian Society for the Advancement of Science," *Impact* (UNESCO), Autumn 1950. See also *Ideas and Opinions*, 357

One may not conclude that the "beginning of the expansion" [of the universe] must mean a singularity in the mathematical sense. All we have to realize is that the [field] equations may not be continued over such regions [of very high density of field and of matter]. This consideration does, however, not alter the fact that the "beginning of the world" really constitutes a beginning, from the point of view of the development of the now existing stars and systems of stars.

From "Appendix for the Second Edition," in *The Meaning of Relativity* (1950), 129

The unified field theory has been put into retirement. It is so difficult to employ mathematically that I have not been able to verify it somehow, in spite of all my efforts. This state of affairs will no doubt last many more years, mostly because physicists have little understanding of logical-philosophical arguments.

To Maurice Solovine, February 12, 1951. Published in *Letters to Solovine*, 123. Einstein Archives 21-277

Science is a wonderful thing if one does not have to earn a living at it. One should earn one's living by work of which one is sure one is capable. Only when we do not have to be accountable to anyone can we find joy in scientific endeavor.

To California student, E. Holzapfel, March 1951. Quoted in Dukas and Hoffmann, *Albert Einstein, the Human Side*, 57. Einstein Archives 59-1013

The betterment of conditions the world over is not strictly dependent on scientific knowledge but on the fulfillment of human traditions and ideals.

To John Cranston, May 16, 1951. Einstein Archives 60-821

One thing I have learned in a long life: that all our science, measured against reality, is primitive and childlike—and yet it is the most precious thing we have.

To Hans Muehsam, July 9, 1951. Einstein Archives 38-408

There is something like a Puritan's restraint in the scientist who seeks truth: he keeps away from everything voluntaristic or emotional.

From the foreword in Philipp Frank, *Relativity: A Richer Truth* (London: Jonathan Cape, 1951), 9. Einstein Archives 1-160

Development of Western science is based on two great achievements: the invention of the formal logical system (in Euclidean geometry) by the Greek philosophers, and the discovery of the possibility of finding out causal relationships by systematic experiment (during the Renaissance).

To J. S. Switzer, April 23, 1953. Einstein Archives 61-381

That no one can make a definite statement about [the unified field theory's] confirmation or nonconfirmation results from the fact that there are no methods of affirming anything with respect to solutions that do not yield to the peculiarities of such a complicated nonlinear system of equations. It is even possible that no one will ever know.

To Maurice Solovine, May 28, 1953. Published in *Letters to Solovine*, 149. Einstein Archives 21-300

In striving to do scientific work, the chance—even for very gifted persons—to achieve something of real value is very small. . . . There is only one way out: devote most of your time to some practical work . . . that agrees with your nature, and spend the rest of it in study. So you will be able . . . to lead a normal and harmonious life even without the special blessings of the Muses.

> To R. Bedi in India who was unsure about what lifework to pursue, July 28, 1953. Quoted in Dukas and Hoffmann, *Albert Einstein, the Human Side*, 59. Einstein Archives 59-180

Now the scientists are bombarding me with questions about my new theory. . . . For two months my colleagues have been jumping on it, each trying to improve on it. But I'm absolutely convinced that it can't be tampered with anymore. I've worked on this theory for a long time to come up with this result.

> Said after his latest equations for a unified field theory were published as an appendix to the fourth edition of *The Meaning of Relativity*. Quoted by Fantova, "Conversations with Einstein," October 16, 1953

*It is possible that there are human emanations of which we are ignorant. You remember how skeptical everyone was about electric currents and invisible waves? Science is still in its infancy.

> In a conversation before 1954. Recalled by Antonina Vallentin, *The Drama of Albert Einstein* (New York: Doubleday, 1954), 155

It is strange that science, which in the old days seemed harmless, should have evolved into a nightmare that causes everyone to tremble.

> To Queen Elisabeth of Belgium, March 28, 1954; quoted in Whitrow, *Einstein*, 89. Einstein Archives 32-410

There is no doubt that the special theory of relativity, if we look at its development in retrospect, was ripe for discovery in 1905.

> To Carl Seelig, February 19, 1955. Einstein Archives 39-069

It appears doubtful that a [classical] field theory can account for the atomistic structure of matter and radiation as well as of quantum phenomena. Most physicists will reply with a firm "no," since they believe that the quantum problem has been solved in principle by other means. However that may be, Lessing's comforting words stay with us: "The struggle for truth is more precious than its assured possession."

> Einstein's final written scientific words, on quantum theory, March 1955, about a month before his death. Written for *Schweizerische Hochschulzeitung;* reprinted in Seelig, *Helle Zeit, dunkle Zeit;* also quoted by Pais in French, *Einstein: A Centenary Volume,* 37. Einstein Archives 1-205

*Since the mathematicians have invaded the theory of relativity, I do not understand it myself anymore.

> Obviously said tongue-in-cheek. Quoted by Carl Seelig, in *Albert Einstein* (Zurich: Europa-Verlag, 1960), 46

When I am judging a theory, I ask myself whether, if I were God, I would have arranged the world in such a way.

> Said to his assistant Banesh Hoffmann. See Harry Woolf, ed., *Some Strangeness in the Proportion* (Reading, Mass.: Addison-Wesley, 1980), 476

Men really devoted to the progress of knowledge concerning the physical world . . . never worked for practical, let alone military, goals.

> Ibid., 510

I have thought a hundred times as much about the quantum problems as I have about general relativity theory.

> As recalled by Otto Stern. Quoted by Pais in French, *Einstein: A Centenary Volume*, 37

I can, if the worse comes to worst, still realize that God may have created a world in which there are no natural laws. In short, chaos. But that there should be statistical laws with definite solutions, i.e., laws that compel God to throw dice in each individual case, I find highly disagreeable.

> As recalled by James Franck. Quoted by C. P. Snow in French, *Einstein: A Centenary Volume*, 6

It follows from the theory of relativity that mass and energy are both different manifestations of the same thing—a somewhat unfamiliar conception for the average man. Furthermore, $E = mc^2$, in which energy is put equal to mass multiplied with the square of the velocity of light, showed that a very small amount of mass may be converted into a very large amount of energy . . . the mass and energy in fact were equivalent.

Read out loud to an audience; filmed and shown in *Nova*'s
Einstein biography on PBS television, 1979

Physics is essentially an intuitive and concrete science. Mathematics is only a means for expressing the laws that govern phenomena.

Quoted by Maurice Solovine in "Introduction" to *Letters to Solovine*, 7–8

An hour sitting with a pretty girl on a park bench passes like a minute, but a minute sitting on a hot stove seems like an hour.

Einstein's explanation of relativity that he gave to his secretary, Helen Dukas, to relay to reporters and other laypersons. Quoted in Sayen, *Einstein in America*, 130

It would be possible to describe everything scientifically, but it would make no sense. It would be a description without meaning—as if you described a Beethoven symphony as a variation of wave pressure.

Quoted in Max Born, *Physik im Wandel meiner Zeit* (Braunschweig: Vieweg, 1966)

A scientist is a mimosa when he himself has made a mistake, and a roaring lion when he discovers a mistake of others.

Quoted in Ehlers, *Liebes Hertz!* 45

On Miscellaneous Subjects

Abortion

A woman should be able to choose to have an abortion up to a certain point in pregnancy.

> To the World League for Sexual Reform, Berlin, September 6, 1929. Quoted in Grüning, *Ein Haus für Albert Einstein*, 305. Einstein Archives 48-304

Achievement

The value of achievement lies in the achieving.

> To D. Liberson, October 28, 1950. Einstein Archives 60-297

Ambition

Nothing truly valuable arises from ambition or from a mere sense of duty; it stems rather from love and devotion toward men and toward objective things.

> To F. S. Wada, an Idaho farmer who requested some words that his son, Albert Wada, could live by as he grew up, July 30, 1947. Quoted in Dukas and Hoffmann, *Albert Einstein, the Human Side*, 46. Einstein Archives 58-934

Animals/Pets

*The love of living creatures is for me the finest and best trait of mankind.

To Valentine Bulgakov, November 4, 1931. Einstein Archives 45-702

Thank you very much for your kind and interesting information. I am sending my heartiest greetings to my namesake, also from our tomcat, who was very interested in the story and even a little jealous. The reason is that his own name, "Tiger," does not express, as in your case, the close kinship to the Einstein family.

To Edward Moses, August 10, 1946, after learning that his ship's crew had rescued a kitten in Germany and named it Einstein. Einstein Archives 57-194

Dr. Dean established that Bibo—the parrot—has a parrot sickness, and that my illness has been due to an infection from him. . . . The poor bird will need thirteen injections—he won't survive them. . . . [Later] Bibo needed only two injections and he is quite happy about it; maybe he'll survive, after all.

Quoted by Fantova, "Conversations with Einstein," February 20 and March 4, 1955. Bibo was a seventy-fifth birthday gift from some admirers the preceding year. He had arrived by post in a box like an ordinary piece of mail, and Einstein took immediate pity on him, for days helping to ease his trauma by trying to cheer him up. The Einsteins had another bird named Bibo, or "Biebchen," while living in Germany.

I know what's wrong, dear fellow, but I don't know how to turn it off.

To his tomcat, Tiger, who seemed depressed because he was housebound due to rain. Recalled by Ernst Straus in his memorial talk, "Albert Einstein, the Man," at UCLA, May 1955, 14–15

The main thing is that *he* knows.

About a friend's dog, Moses, whose long fur made it difficult to tell one end from the other. From an interview, January 15, 1979, with Margot Einstein by J. Sayen; quoted in Sayen, *Einstein in America*, 131

The dog is very smart. He feels sorry for me because I receive so much mail; that's why he tries to bite the mailman.

Regarding his dog, Chico. Quoted in Ehlers, *Liebes Hertz!* 162

Art and Science

Where the world ceases to be the scene of our personal hopes and wishes, where we face it as free beings, admiring, questioning, and observing, there we enter the realm of art and science. We do science when we reconstruct in the language of logic what we have seen and experienced; we do art when we communicate through forms whose connections are not accessible to the conscious mind yet we intuitively recognize them as something meaningful.

For a magazine on modern art, *Menschen. Zeitschrift neuer Kunst* 4 (February 1921), 19. See also *CPAE*, Vol. 7, Doc. 51

Astrology

The reader should note [Kepler's] remarks on astrology. They show that the inner enemy, conquered and rendered innocuous, was not yet completely dead.

From "Introduction" in *Johannes Kepler: Life and Letters* by Carola Baumgardt (New York: Philosophical Library, 1951). See also "Attributed to Einstein" at the back of the book, with a source that refutes the allegation that Einstein believed in astrology.

Birth Control

I am convinced that some political and social activities and practices of the Catholic organizations are detrimental and even dangerous for the community as a whole, here and everywhere. I mention here only the fight against birth control at a time when overpopulation in various countries has become a serious threat to the health of people and a grave obstacle to any attempt to organize peace on this planet.

To a reader of the Brooklyn *Tablet*, the newspaper of the diocese of Brooklyn and Queens, 1954, who questioned Einstein about whether he had been correctly quoted on the subject

Birthdays

My dear little sweetheart . . . first, my belated cordial congratulations on your birthday yesterday, which I had forgotten once again.

To girlfriend Mileva Marić, his future wife, December 19, 1901. *CPAE*, Vol. 1, Doc. 130

My birthday affords me the welcome opportunity to express my feelings of deep gratitude for the ideal working and living conditions which have been placed at my disposal in the United States.

From a statement issued on his sixtieth birthday. *Science* 89, n.s. (1939), 242

What is there to celebrate? Birthdays are automatic things. Anyway, birthdays are for children.

From an interview, *New York Times*, March 12, 1944

My birthday was a natural disaster, a shower of paper, full of flattery, under which one almost drowned.

To Hans Muehsam, March 30, 1954, on Einstein's seventy-fifth birthday. Einstein Archives 38-434

Books

What I have to say about this book can be found inside the book.

Reply to a *New York Times* reporter's request for a comment on Einstein's book *The Evolution of Physics*, written with Leopold Infeld. Quoted in Ehlers, *Liebes Hertz!* 65

I am reading Dostoyevsky (*The Brothers Karamazov*). It's the most wonderful thing that has ever fallen into my hands.

To Heinrich Zangger, March 26, 1920. *CPAE*, Vol. 9, Doc. 361

I am in raptures about *The Brothers Karamazov*. It is the most wonderful book I have ever put my hands on.

To Paul Ehrenfest, April 7, 1920. *CPAE*, Vol. 9, Doc. 371

Causality

The causal way of looking at things always answers only the question, "Why?" but never "To what end?" . . . However, if someone asks, "For what purpose should we help one another, make life easier for each other, make beautiful music together, have inspired thoughts?" he would have to be told, "If you don't feel the reasons, no one can explain them to you." Without this primary feeling we are nothing and had better not live at all.

To Hedwig Born, August 31, 1919. *CPAE*, Vol. 9, Doc. 97

I believe that whatever we do or live for has its causality. It is good, however, that we do not know what it is.

From a conversation with Indian mystic, poet, and musician Rabindranath Tagore in Berlin, August 19, 1930. Published in *Asia* 31 (March 1931)

China and the Chinese

*I would imagine life among the Chinese to be actually quite fine and attractive. I found the few exemplars I met extraordinarily appealing. From the human point of view these people, in their finely proportioned shapes, actually seem to be far superior to us.

To Franz Rusch, who was teaching in Tientsin and felt lonely, March 18, 1921. *CPAE*, Vol. 12, Doc. 105

*As to Chinese youths, I believe that they are bound to make great contributions to science in the future.

From an address at a reception in Shanghai at the residence of Wang Yiting, November 13, 1922. Quoted in Hu, *China and Albert Einstein*, 72

*Outwardly, the Chinese attract attention by their industry, the small demands of their lifestyles, and their abundance of children. . . . Yet to a large extent they are a burdened people, cracking stones day after day and carrying them for pennies. They seem to be too stolid to recognize the horror of their lot. . . . In Shanghai, the Europeans are the master class and the Chinese their servants. . . . They appear not in any way connected to their great intellectual history. As good-natured laborers they are appreciated by the Europeans, to whom as such they are far inferior intellectually.

From his Travel Diary, December 31, 1922, and January 1, 1923. Einstein Archives 29-131

Christmas

*Christmas is the festival of peace. Every year it comes in its own good time. But peace inside and among us can come only through persistent effort. This holiday reminds us that all people yearn for peace. Every year it admonishes us to be vigilant against the enemies of peace that lurk inside all of us, lest they cause harm not only at Christmastime but throughout the year.

From a Christmas message for an international radio broadcast, November 28, 1948. Einstein Archives 28-850

Clarity

All my life I have been a friend of well-chosen, sober words and of concise presentation. Pompous phrases and words give me goose bumps whether they deal with the theory of relativity or with anything else.

Quoted in *Berliner Tageblatt*, August 27, 1920, 1–2. See also *CPAE*, Vol. 7, Doc. 45

Class

The distinctions separating the social classes are false, in the last analysis they rest on force.

From "What I Believe," *Forum and Century* 84 (1930), 193–194

Clothes

If I were to start taking care of my grooming, I would no longer be my own self. . . . So the hell with it. If you find me so repulsive, then look for a boyfriend who is more appealing to female tastes. But I will continue to be unconcerned about it, which surely has the advantage that I'm left in peace by many a fop who would otherwise come to see me.

To future second wife, Elsa Löwenthal, ca. December 2, 1913. *CPAE*, Vol. 5, Doc. 489

Only a certain regimen regarding attire, etc., so as not to be counted among the rejects of the local human race, disturbs my peace of mind somewhat.

To the Hurwitz family, May 4, 1914, on his new life in Berlin. *CPAE*, Vol. 8, Doc. 6

*If they want to see me, here I am. If they want to see my clothes, open my closet.

Said to Elsa after she suggested he change his clothes before receiving a delegation of visitors from German president von Hindenburg, 1932. Quoted in Brian, *Einstein, a Life*, 235

I like neither new clothes nor new kinds of food.

Quoted in Pais, *Subtle Is the Lord*, 16

"Why should I? Everyone knows me there" (upon being told by his wife to dress properly when going to the office). "Why should I? No one knows me there" (upon being told to dress properly for his first big conference).

Quoted in Ehlers, *Liebes Hertz!* 87

I have reached an age when, if someone tells me to wear socks, I don't have to.

Quoted by neighbor and fellow physicist Allen Shenstone, in Sayen, *Einstein in America*, 69

When I was young I found out that the big toe always ends up making a hole in a sock. So I stopped wearing socks.

Recalled by Philippe Halsman, 1947. Quoted in French, *Einstein: A Centenary Volume*, 27

Competition

I no longer need to take part in the competition of the big brains. Participating [in the process] has always seemed to me to be an awful type of slavery no less evil than the passion for money or power.

To Paul Ehrenfest, May 5, 1927, regarding the rat race for academic promotions. Quoted in Dukas and Hoffmann, *Albert Einstein, the Human Side*, 60. Einstein Archives 10-163

Comprehensibility

The eternal mystery of the world is its comprehensibility. . . . The fact that it is comprehensible is a miracle.

From "Physics and Reality," *Journal of the Franklin Institute* 221, no. 3 (March 1936), 349–382. Reprinted in *Ideas and Opinions*, 292. Popularly rephrased as, "The most incomprehensible thing about the universe is that it is comprehensible."

Compromise

The path to a lazy compromise is a one-way street. There is no U-turn and no stopping.

To Johanna Fantova, October 9, 1948, one of three aphorisms sent to her. Einstein Archives 87-347

Conscience

Never do anything against conscience even if the state demands it.

From a conversation with Virgil G. Hinshaw, Jr., quoted in
Hinshaw's contribution to Schilpp, *Albert Einstein: Philosopher Scientist* (1949), 653

*Conscience supersedes the authority of the law of
the state.

Another version of the above statement. From "Human
Rights," a message to the Chicago Decalogue Society of
Lawyers upon receiving its award for contributions to
human rights. The message had been written just before
December 5, 1953 (Einstein Archives 28-1012), and was
translated and recorded before being played at the cere-
mony on February 20, 1954. See Rowe and Schulmann,
Einstein on Politics, 497

Creativity

Without creative personalities able to think and
judge independently, the upward development of
society is as unthinkable as the development of the
individual personality without the nourishing soil
of the community.

From "Society and Personality," 1932. Published in *Mein
Weltbild* (1934), 12; reprinted in *Ideas and Opinions*, 14

I lived in solitude in the country and noticed how
the monotony of a quiet life stimulates the creative
mind.

From a speech, "Science and Civilization," at the Royal Albert Hall, London, October 3, 1933. Published in 1934 as
"Europe's Danger–Europe's Hope." Quoted in *The Times*

(London), October 4, 1933, 14, though the remarks are not in the original written version of the speech. Einstein Archives 28-253

Crises

Only through perils and upheavals can Nations be brought to further developments. May the present upheavals lead to a better world.

From a speech, "Science and Civilization," at the Royal Albert Hall, London, October 3, 1933. Published in 1934 as "Europe's Danger–Europe's Hope." Quoted in *The Times* (London), October 4, 1933, 14. Einstein Archives 28-253

Curiosity

The important thing is not to stop questioning. Curiosity has its own reason for existing. One cannot help but be in awe when one contemplates the mysteries of eternity, of life, of the marvelous structure of reality. It is enough if one tries to comprehend only a little of this mystery every day.

From the memoirs of William Miller, an editor, quoted in *Life* magazine, May 2, 1955

This delicate little plant [curiosity], aside from stimulation, stands mainly in need of freedom.

Written in 1946 for "Autobiographical Notes," 17

Death Penalty

I have reached the conviction that the abolition of the death penalty is desirable. Reasons: (1) Irreversibility in the event of an error in justice; (2) detrimental moral influence on those who . . . have to carry out the procedure.

> To a Berlin publisher, November 3, 1927. Einstein Archives 46-009. However, several months earlier, according to the *New York Times*, the story had been a bit different: "Professor Einstein does not favor the abolition of the death penalty. . . . He could not see why society should not rid itself of individuals proved socially harmful. He added that society had no greater right to condemn a person to life imprisonment than it had to sentence him to death"; see the *New York Times*, March 6, 1927; also noted in Pais, *Einstein Lived Here*, 174

I am not for punishment at all, but only for measures that save society and protect it. In principle, I would not be opposed to killing individuals who are worthless or dangerous in that sense. I am against it only because I do not trust people, i.e., the courts. What I value in life is quality rather than quantity.

> To Valentine Bulgakov, who had been Tolstoy's secretary, who asked Einstein his thoughts on war and the death penalty, November 4, 1931. Einstein Archives 45-702

Doctors

*Oh, zose surgeons, zey are vizards.

> Encouragement to neighbor Eric Rogers, who was about
> to undergo an operation. Related by Ralph Baierlein, who
> later taught with Rogers at Princeton (letter to me of
> March 21, 2006)

England, the English, and the English Language

*If scholars would take their profession more seri-
ously than their political passions, they would guide
their actions more according toward cultural factors
than political ones. . . . In this regard, the English
have behaved far more nobly than our colleagues
here. . . . How magnificent their attittude has been
toward me and relativity theory! . . . I can only say:
Hats off to the fellows!

> To Fritz Haber, March 9, 1921. Einstein felt that relativity
> theory had become a political issue in Germany. *CPAE*,
> Vol. 12, Doc. 88

*The wonderful experiences in England are still
fresh in my mind and like a dream. The impression
that this land and its wonderful intellectual and political traditions has made on me was even deeper,
longer lasting, and greater than I had anticipated.

> To Lord Richard B. S. Haldane, June 21, 1921. *CPAE*,
> Vol. 12, Doc. 155

Whereas in Germany, in general, the judgment of my theory depended upon the political orientation of the newspapers, the English scientists' attitude has proven that their sense of objectivity cannot be muddled by political viewpoints.

From "How I Became a Zionist," *Jüdische Rundschau*, June 21, 1921. See *CPAE*, Vol. 7, Doc. 56

More than any other people, you Englishmen have carefully cultivated the bond of tradition and preserved the living and conscious continuity of succeeding generations. You have in this way endowed with vitality and reality the distinctive soul of your people and the soaring soul of humanity.

To the Royal Society (U.K.) on the occasion of the Newton bicentenary, March 1927. Reprinted in *Nature* 119 (1927), 467. Einstein Archives 1-058

I cannot write in English, because of the treacherous spelling. When I am reading, I only hear it and am unable to remember what the written word looks like.

To Max Born, September 7, 1944, expressing the difficulty he had with the language of his adopted country, even though he was eager to become an American citizen. In Born, *Born-Einstein Letters*, 145. Einstein Archives 8-207

Epistemology

When I think of the most able students I have encountered in my teaching—I mean those who have distinguished themselves not only by skill but by independence of thought—then I must confess that all have had a lively interest in epistemology. No one can deny that epistemologists have paved the road for progress [toward the theory of relativity]; Hume and Mach, at least, have helped me considerably, both directly and indirectly.

From "Ernst Mach," *Physikalische Zeitschrift* 17 (1916). *CPAE*, Vol. 6, Doc. 29

Epistemology without contact with science becomes an empty scheme. Science without epistemology is—insofar as it is thinkable at all—primitive and muddled.

From "Reply to Criticisms," in Schilpp, *Albert Einstein: Philosopher-Scientist*, 684

*In thinking, we use with a certain "right," concepts to which there is no access from the materials of sensory experience if the situation is viewed from a logical point of view.

From "Russell's Theory of Knowledge," in Paul Schilpp, ed., *The Philosophy of Bertrand Russell* (Library of Living Philosophers, 1944), 287. Einstein Archives 1-139

Flying Saucers and Extraterrestrials

*There is every reason to believe that Mars and other planets are inhabited. But if intelligent creatures do exist, as we may assume they do elsewhere in the universe, I should not expect them to try to communicate with the earth by wireless. Light rays, the direction of which can be controlled much more easily, would more probably be the first method attempted.

From an interview in Berlin on the "Mystic Wireless." Reported in the *Daily Mail* (London), January 31, 1921, 5. *CPAE*, Vol. 12, Calendar

I have no reason to believe that there is something real behind the stories of the "Flying Saucers."

To a boy in Hartford, Connecticut, November 15, 1950. Einstein Archives 59-510. Einstein believed that people should not read science fiction—that it distorts science and gives people the *illusion* of understanding science.

Those people have seen *something*. What it is, I do not know and I am not curious to know.

To L. Gardner, July 23, 1952. Einstein Archives 59-803

Force

Where belief in the omnipotence of physical force gets the upper hand in political life, this force takes on a life of its own and proves stronger than the men who think about using force as a tool.

From an address at Carnegie Hall in New York on receiving the One World Award, April 27, 1948. Published in *Out of My Later Years*; reprinted in *Ideas and Opinions*, 147

Games

I do not play games. . . . There is no time for it. When I get through with work, I don't want anything that requires the working of the mind.

Quoted in the *New York Times*, March 28, 1936, 34: 2. However, Einstein did like to play with puzzles, though this hobby may have started later in his life.

Good Acts

Good acts are like good poems. One may easily get their drift, but they are not always rationally understood.

To Maurice Solovine, April 9, 1947. Published in *Letters to Solovine*, 99, 101. Einstein Archives 21-250

Graphology

It was interesting to me that it is possible to classify handwriting in such an exhaustive manner. I also appreciate that you were able to separate clearly the objective characteristics from the purely intuitive, which, by the way, should not be discredited that much by the example of Hitler.

Handwritten letter to graphologist Thea Lewinson, September 4, 1942 (for sale on eBay, November 5, 2003)

Home

It is not so important where one settles down. The best thing is to follow your instincts without too much reflection.

To Max Born, March 3, 1920. In Born, *Born-Einstein Letters*, 25. *CPAE*, Vol. 9, Doc. 337

Homosexuality

Homosexuality should not be punishable except to protect children.

To the World League for Sexual Reform, Berlin, September 6, 1929. Quoted in Grüning, *Ein Haus für Albert Einstein*, 305–306. Einstein Archives 48-304

Immigrants

*[Immigrants] have contributed in their way to the flowering of the community, and their individual striving and suffering have remained unknown.

Said at the dedication of the Wall of Fame at the World's Fair in New York, 1939–1940. Einstein Archives 28-529

*Unemployment is *not* decreased by restricting immigration. For [unemployment] depends on faulty distribution of work among those capable of work.

Immigration increases consumption as much as it does demand on labor. Immigration strengthens not only the internal economy of a sparsely populated country, but also its defensive power.

Ibid.

Individuals/Individuality

What is truly valuable in our bustle of life is not the nation . . . but the creative and impressionable individuality, the personality—he who produces the noble and sublime while the common herd remains dull in thought and insensible in feeling.

From "What I Believe," *Forum and Century* 84 (1930), 193–194. See Rowe and Schulmann, *Einstein on Politics*, 229

Everyone should be respected as an individual, but not idolized.

Ibid.

Valuable achievement can sprout from human society only when it is sufficiently loosened to make possible the free development of an individual's abilities.

From an article on tolerance, June 1934. Einstein Archives 28-280

*It is not possible to imagine a forest made up only of vines. It needs trees that are able to stand by virtue of their own strength.

> To Johanna Fantova, October 9, 1948, one of three aphorisms sent to her. Einstein Archives 87-347

While it is true that an inherently free and scrupulous person may be destroyed, such an individual can never be enslaved or used as a blind tool.

> From "On the Moral Obligation of the Scientist," for the Italian Society for the Advancement of Science, October 1950. Einstein Archives 28-882

It is important for the common good to foster individuality: for only the individual can produce the new ideas which the community needs for its continuous improvement and requirements—indeed, to avoid sterility and petrification.

> From a message for a Ben Schemen dinner, March 1952. Einstein Archives 28-931

Intelligence

It is abhorrent to me when a fine intelligence is paired with an unsavory character.

> To Jakob Laub, May 19, 1909. *CPAE*, Vol. 5, Doc. 161

We have been endowed with just enough intelligence to be able to see clearly just how utterly inadequate that intelligence is when we are confronted with what exists. If this humility could be imparted to everybody, the world of human endeavors would become more appealing.

> To Queen Elisabeth of Belgium, September 19, 1932. Quoted in Grüning, *Ein Haus für Albert Einstein*, 305. Einstein Archives 32-353

We should take care not to make intellect our god; it has, of course, powerful muscles, but no personality.

> From "The Goal of Human Existence," April 11, 1943. Published in *Out of My Later Years*, 235. Einstein Archives 28-587

Intuition

All great achievements of science must start from intuitive knowledge, namely, in axioms, from which deductions are then made. . . . Intuition is the necessary condition for the discovery of such axioms.

> 1920. Quoted by Moszkowski, *Conversations with Einstein*, 180

I believe in intuitions and inspirations. . . . I sometimes *feel* that I am right. I do not *know* that I am.

> From an interview with G. S. Viereck, "What Life Means to Einstein," *Saturday Evening Post*, October 26, 1929; reprinted in Viereck, *Glimpses of the Great*, 446

Invention

Invention is not the product of logical thought, even though the final product is tied to a logical structure.

Written for *Schweizerische Hochschulzeitung*, 1955. Reprinted in Seelig, *Helle Zeit, Dunkle Zeit.* Quoted in Pais, *Subtle Is the Lord,* 131. Einstein Archives 1-205

Italy and the Italians

*On the 15th [of October] I'll leave with my son for Bologna and show off my sauerkraut-Italian. The grandchildren of Dante are really in store for something!

To Hermann Anschütz-Kaempfe, October 11, 1921. *CPAE*, Vol. 12, Doc. 263

The ordinary Italian . . . uses words and expressions of a high level of thought and cultural content. . . . The people of northern Italy are the most civilized people I have ever met.

Quoted by H. Cohen in *Jewish Spectator*, January 1969, 16

The happy months of my sojourn in Italy are my most beautiful memories.

To Ernesta Marangoni, August 16, 1946. In *Physis* 18 (1976), 174–178. Einstein Archives 57-113

Japan and the Japanese

*I well remember your visit to Bern, especially since you were the first Japanese, indeed the first East Asian, whose acquaintance I ever made. You amazed me with your great theoretical knowledge.

> To Ayao Kuwaki, December 38, 1920. Einstein had met Kuwaki in March 1909. *CPAE*, Vol. 10, Doc. 246

*The invitation to Tokyo pleased me very much, especially considering my long-standing interest in the people and cultures of East Asia.

> To Koshin Murobuse, September 27, 1921, before Einstein embarked on his travels in 1922. *CPAE*, Vol. 12, Doc. 246

The Japanese loves his country and its people more than others do . . . yet he feels like a stranger in foreign countries more than others. I have learned . . . to understand the shyness of the Japanese toward Europeans and Americans: in our countries, education is focused entirely on struggling to survive as individuals. . . . Family bonds are weakened, and . . . isolation of the individual is looked upon as a necessary consequence in the struggle for existence. . . . It is completely different in Japan. The individual here is left to himself much less than in Europe and America. Public opinion here is even greater than in our countries, and sees to it that family structure is not weakened.

Kaizo 5, no. 1 (January 1923), 339. In November–December 1922 Einstein went on a six-week trip to Japan, where he was received with great enthusiasm. In addition to Einstein's other attributes, its citizens may also have been curious about him because the Japanese characters for "relativity principle" are very similar to those for "love" and "sex" (see Fölsing, *Albert Einstein*, 528). Einstein Archives 36-477.1

This flowerlike being: here the ordinary mortal must defer to the poet's words.

On Japanese women. Ibid.

May they not forget to keep pure the great heritage that puts them ahead of the West: the artistic configuration of life, the simplicity and modesty of personal needs, and the purity and serenity of the Japanese soul.

Ibid., 338

It was wonderful in Japan—genteel manners, a lively interest in everything, an artistic sense, intellectual honesty together with common sense—a wonderful people in a picturesque land.

To Maurice Solovine, May 20, 1923. Published in *Letters to Solovine*, 58–59. Einstein Archives 21-189

I have, for the first time, seen a happy and healthy society whose members are fully absorbed in it.

To Michele Besso, May 24, 1924. Einstein Archives 7-349

Japan is now like a great kettle without a safety valve. It does not have enough land to enable its population to exist and develop. The situation must somehow be remedied if we are to avoid a terrible conflict.

> Quoted in the *New York Times*, May 17, 1925, in an interview with Herman Bernstein. Quoted in Nathan and Norden, *Einstein on Peace*, 75. Three years later, the Japanese occupied the Shantung region of China and many years of conflict ensued.

Knowledge

*Knowledge exists in two forms—lifeless, stored in books, and alive in the consciousness of men. The second form of existence is after all the essential one; the first, indispensable as it may be, occupies only an inferior position.

> In the last paragraph of "Message in Honor of Morris Raphael Cohen" for the Cohen Student Memorial Fund, November 15, 1949. Reprinted in *Ideas and Opinions*, 80

Love

Love brings much happiness, much more so than pining for someone brings pain.

> To Marie Winteler, his first girlfriend, April 21, 1896 (at age seventeen). *CPAE*, Vol. 1, Doc. 18

Falling in love is not at all the most stupid thing that people do—but gravitation cannot be held responsible for it.

> A scribbled response on a letter from Frank Wall, 1933, in reply to his letter asking if it would be "reasonable to assume that it is while a person is standing on his head [i.e., as the world rotates]—or rather upside down—he falls in love and does other stupid things." Quoted in Dukas and Hoffmann, *Albert Einstein, the Human Side*, 56. Einstein Archives 31-845

Where there is love, there is no imposition.

> To editor and friend Saxe Commins, Summer 1953. Quoted in Sayen, *Einstein in America*, 294

I am sorry that you are having difficulties bringing your girlfriend [from Dublin to the United States]. But as long as she is there and you are here, you should be able to maintain a harmonious relationship. So why do you want to press the issue?

> To Cornel Lanczos, February 14, 1955. Einstein Archives 15-328

Marriage

My parents . . . think of a wife as a man's luxury that he can afford only when he is making a comfortable living. I have a low opinion of this view of the relationship between man and wife, because it makes the wife and the prostitute distinguishable only insofar as the former is able to secure a lifelong contract from the man because of her favorable social rank.

> To Mileva Marić, August 6, 1900. See *The Love Letters*, 23. *CPAE*, Vol. 1, Doc. 70

It is not a lack of real affection that scares me away again and again from marriage. Is it a fear of the comfortable life, of nice furniture, of dishonor that I burden myself with, or even the fear of becoming a contented bourgeois?

> To Elsa Löwenthal, after August 3, 1914. *CPAE*, Vol. 8, Doc. 32

The solitude and peace of mind are serving me quite well, not the least of which is due to the excellent and truly enjoyable relationship with my cousin; its stability will be guaranteed by the avoidance of marriage.

> To Michele Besso, February 12, 1915. *CPAE*, Vol. 8, Doc. 56. Of course, Einstein did marry Elsa four years later.

My aim is to smoke it, but as a result things tend to get clogged up, I'm afraid. Life, too, is like smoking, especially marriage.

> Recalled by Ippei Okamoto, a Japanese cartoonist, while Einstein was in Japan in 1922. Okamato had asked him if he smoked his pipe for the pleasure of smoking, or simply to engage in unclogging and refilling his pipe. Quoted in Kantha, *An Einstein Dictionary*, 199; and *American Journal of Physics* 49 (1981), 930–940

Why should one not admit a man [to the United States] . . . who dares to oppose every war except the inevitable one with his own wife?

In a reply to the right-wing Women Patriot Corporation, who felt Einstein would be a bad influence on Americans and should not be allowed to enter the country, December 1932. Einstein Archives 28-213

Marriage is the unsuccessful attempt to make something lasting out of an incident.

Quoted by Otto Nathan, April 10, 1982, in an interview with J. Sayen for *Einstein in America*, 80. Also mentioned in Fantova, "Conversations with Einstein," December 5, 1953

That is dangerous—but then, *any* marriage is dangerous.

In answer to the question of a Jewish student at Princeton about whether interfaith marriage should be tolerated. Sayen, *Einstein in America*, 70

Marriage is but slavery made to appear civilized.

Quoted by Konrad Wachsmann in Grüning, *Ein Haus für Albert Einstein*, 159

Marriage makes people treat each other as articles of property and no longer as free human beings.

Ibid.

Materialism

Human beings can attain a worthy and harmonious life only if they are able to rid themselves, within the limits of human nature, of striving to fulfill wishes of the material kind. The goal is to raise the spiritual values of society.

> At a planning conference in Princeton of American Friends of Hebrew University, September 19, 1954. Quoted in the *New York Times*, September 20, 1954. Einstein Archives 37-354

Miracles

I admit that thoughts influence the body.

> Quoted by W. Hermanns in *A Talk with Einstein*, October 1943. Einstein Archives 55-285

A "miracle" is an exception from lawfulness; hence, where lawfulness does not exist, its exception, i.e., a miracle, also cannot exist.

> Recalled by David Reichinstein in *Die Religion des Gebildeten* (Zurich, 1941), 21. Quoted and discussed in Jammer, *Einstein and Religion*, 89

Morality

One must shy away from questionable undertakings, even when they bear a high-sounding name.

To Maurice Solovine, May 20, 1923, on Einstein's resignation from a League of Nations commission. Published in
Letters to Solovine, 59. Einstein Archives 21-189

Morality is of the highest importance—but for us,
not for God.

To M. Schayer, August 1, 1927. Quoted in Dukas and Hoffmann, *Albert Einstein, the Human Side*, 66. Einstein Archives
48-380

The content of scientific theory itself offers no moral
foundation for the personal conduct of life.

From "Science and God: A Dialogue," *Forum and Century*
83 (June 1930), 373

The destiny of civilized humanity depends more
than ever on the moral forces it is capable of
generating.

From "Address to the Student Disarmament Meeting,"
February 27, 1932. Published in *Mein Weltbild*; reprinted in
Ideas and Opinions, 94, and *New York Times*, February 28,
1932

There is nothing divine about morality; it is a purely
human affair.

From "The Religious Spirit of Science." Published in *Mein
Weltbild* (1934), 18; reprinted in *Ideas and Opinions*, 40

Let us not forget that human knowledge and skills
alone cannot lead humanity to a happy and digni-

fied life. Humanity has every reason to place the proclaimers of high moral standards and values above the discoverers of objective truth. What humanity owes to personalities like Buddha, Moses, and Jesus ranks for me higher than all the achievements of the inquiring constructive mind.

Statement in September 1937 for a UNIDFNT "Preaching Mission." Quoted in Dukas and Hoffmann, *Albert Einstein, the Human Side*, 70. Einstein Archives 28-401

Morality is not a fixed and stark system. . . . It is a task never finished, something that is always present to guide our judgment and inspire our conduct.

From "Morals and Emotions," a commencement address at Swarthmore College, Pennsylvania, June 6, 1938. Quoted in the *New York Times*, June 7, 1938. Einstein Archives 29-083

The most important human endeavor is the striving for morality in our actions. Our inner balance and even our very existence depend on it. Only morality in our actions can give beauty and dignity to life.

To the Reverend C. Greenway, a minister in Brooklyn, November 20, 1950. Quoted in Dukas and Hoffmann, *Albert Einstein, the Human Side*, 95. Einstein Archives 28-894, 59-871

Without "ethical culture," there is no salvation for humanity.

From "The Need for Ethical Culture," January 5, 1951. Einstein Archives 28-904

Mysticism

*The mystical trend of our present time, especially evident in the enthusiastic growth of so-called theosophy and spiritualism, is to me a symptom of confusion and weakness. Since our inner experiences consist of reproductions and combinations of sensory impressions, the concept of a soul without a body seems to me to be empty and devoid of meaning.

> To Viennese poet Lili Halpern-Neuda, February 5, 1921. *CPAE*, Vol. 12, Doc. 41. See Doc. 40, n. 3, for background.

I have never imputed to Nature a purpose or a goal, or anything that could be understood as anthropomorphic. What I see in Nature is a magnificent structure that we can comprehend only very imperfectly, and that must fill a thinking person with a feeling of humility. This is a genuinely religious feeling that has nothing to do with mysticism.

> To Ugo Onofri, 1954 or 1955. Quoted in Dukas and Hoffmann, *Albert Einstein, the Human Side*, 39. Einstein Archives 60-758

Nature

The most beautiful gift of nature is that it gives one pleasure to look around and try to comprehend what we see.

Aphorism, February 23, 1953. In *Essays Presented to Leo Baeck on the Occasion of His Eightieth Birthday* (London: East and West Library, 1954). Einstein Archives 28-962

Pipe Smoking

Pipe smoking contributes to a somewhat calm and objective judgment in all human affairs.

To Montreal Pipe Smokers Club, upon acceptance of life membership, March 7, 1950. Quoted in the *New York Times*, March 12, 1950. Einstein Archives 60-125. Einstein was said to be so fond of his pipe that he held on to it even after he fell into the water during a boating accident; see Ehlers, *Liebes Hertz!*, 149. *See also* under "Marriage"

Posterity

Dear Posterity: If you have not become more just, more peaceful, and in general more sensible than we are (or were), then may the Devil take you! Respectfully expressing his opinion with this devout hope is (or was) your Albert Einstein.

Princeton, May 4, 1936. Message to posterity written on parchment. It was put in an airtight metal box in the cornerstone of the Schuster publishing house (today Simon & Schuster) in New York. Einstein Archives 51-798

The Press

The Press, which is mostly controlled by vested interests, has an excessive influence on public opinion.

> From "Impressions of the U.S.A," ca. 1931. Source misquoted in *Ideas and Opinions*, 5. Einstein Archives 28-167

Prohibition

Nothing is more destructive of respect for the government and the law of the land than passing laws that cannot be enforced. It is an open secret that the dangerous increase in crime in this country is closely connected with this.

> Ibid.

I don't drink, so I couldn't care less.

> Statement about Prohibition at a press conference on his arrival in San Diego, December 30, 1930, on board the *Oakland*. Shown in *Nova*'s Einstein biography on PBS, 1979, and in A&E Television's Einstein biography, VPI International, 1991. Einstein disliked alcohol and remained a teetotaler in his later years, perhaps due to his sensitive digestive system; see Fölsing, *Albert Einstein*, 81

Psychoanalysis

I should very much like to remain in the darkness of not having been analyzed.

To German psychotherapist H. Freund, in response to the request that he take part in a study based on Adlerian psychology, January 1927. Quoted in Dukas and Hoffmann, *Albert Einstein, the Human Side*, 35. Einstein Archives 46-304

I am not able to venture a judgment on so important a phase of modern thought. However, it seems to me that psychoanalysis is not always salutary. It may not always be helpful to delve into the unconscious.

From an interview with G. S. Viereck, "What Life Means to Einstein," *Saturday Evening Post*, October 26, 1929; reprinted in Viereck, *Glimpses of the Great*, 442

I don't think it's impossible that dreams are suppressed wishes, but I'm not convinced.

Quoted in Fantova, "Conversations with Einstein," November 5, 1953

Public Speaking

I have just got a new theory of eternity.

Alleged remark to a table-mate while listening to long-winded speeches at a National Academy of Science dinner honoring him. Cited by Daniel Greenberg, "A Statue without Stature," *Washington Post*, December 12, 1978

Rickshaw Pullers

I felt extremely ashamed to be part of such hideous treatment of human beings but couldn't do anything about it. . . . They know how to beseech and beg every tourist until he capitulates.

> From his Travel Diary, October 28, 1922, Colombo, Ceylon (now Sri Lanka). Einstein stopped there en route to Singapore, Hong Kong, Shanghai in China, and Japan.

Sailing

Sailing in the secluded coves of the coast here is more than relaxing. . . . I have a compass that shines in the dark, like a serious seafarer. But I am not so talented in this art, and I am satisfied if I can manage to get myself off the sandbanks on which I become lodged.

> To Queen Elisabeth of Belgium, March 20, 1954. Einstein Archives 32-385

The sport that demands the least energy.

> Quoted by A. P. French, in French, *Einstein: A Centenary Volume*, 61

Sculpture

The ability to portray people in motion requires the highest measure of intuition and talent.

> Quoted by Konrad Wachsmann in Grüning, *Ein Haus für Albert Einstein*, 240. Einstein's stepdaughter, Margot, was a sculptor, and Einstein himself sat for a number of sculptures.

Sex Education

Regarding sex education: no secrets!

> To the World League for Sexual Reform, Berlin, September 6, 1929. Einstein Archives 48-304

Success

Try to become not a man of success, but try rather to become a man of value.

> Quoted by William Miller in *Life* magazine, May 2, 1955

Thinking

The words of the language, as they are written or spoken, do not seem to play any role in my mechanism of thought.

> To Jacques Hadamard, June 17, 1944. Quoted in *An Essay on the Psychology of Invention in the Mathematical Field*, Appendix 2. Einstein Archives 12-056

I vill a little t'ink.

> According to Banesh Hoffmann, this is the phrase Einstein used in his broken English when he needed more time to think about a problem. Quoted in French, *Einstein: A Centenary Volume*, 153; Hoffmann, *Albert Einstein, Creator and Rebel*, 231, quotes Leopold Infeld saying the same about Einstein. Also related to me by Einstein's biographer Abraham Pais, who knew Einstein.

[I have no doubt] that our thinking goes on for the most part without use of signs (words), and beyond that to a considerable degree unconsciously. For how, otherwise, should it happen that sometimes we "wonder" quite spontaneously about some experience? This "wondering" seems to occur when an experience comes into conflict with a world of concepts which is already sufficiently fixed within us. . . . The development of this thought world is in a certain sense a continual flight from "wonder."

Written in 1946 for "Autobiographical Notes," 8–9

Truth

The effort to strive for truth has to precede all other efforts.

To Alfredo Rocco, November 16, 1931. Einstein Archives 34-725

The search for truth and knowledge is one of the finest attributes of man—though often it is most loudly voiced by those who strive for it the least.

From "The Goal of Human Existence," broadcast for the United Jewish Appeal, April 11, 1943. Einstein Archives 28-587

Truth is what stands the test of experience.

Closing words of foreword to Philipp Frank, *Relativity: A Richer Truth* (Boston: Beacon Press, 1950). Quoted in *Out of My Later Years*, rev. ed., 115. Einstein Archives 1-160

It is difficult to say what truth is, but sometimes it is so easy to recognize a falsehood.

> To Jeremiah McGuire, October 24, 1953. Einstein Archives 60-483

Whoever is careless with truth in small matters cannot be trusted in important affairs.

> From a draft of a television address to be delivered on occasion of the seventh anniversary of Israel's independence. Written in April 1955, about a week before Einstein's death. Quoted in Nathan and Norden, *Einstein on Peace*, 640. Einstein Archives 60-003

The pompous vocabulary and revolutionary overtones of your letter make me suspicious. Truth tends to present itself modestly and in simple garb.

> To Hans Wittig, May 3, 1920. Einstein Archives 45-274

Vegetarianism

Although I have been prevented by outward circumstances from observing a strictly vegetarian diet, I have long been an adherent to the cause in principle. Besides agreeing with the aims of vegetarianism for aesthetic and moral reasons, it is my view that a vegetarian manner of living by its purely physical effect on the human temperament would most beneficially influence the lot of mankind.

> To Hermann Huth, December 27, 1930. Supposedly published in the German magazine *Vegetarische Warte,* which

was published from 1882 to 1935; Huth was the vice-president of the Vegetarier-Bund, the German vegetarian society that published the magazine. (Thanks to David Hurwitz for this information.) Einstein Archives 46-756

I have always eaten animal flesh with a somewhat guilty conscience.

To Max Kariel, August 3, 1953. Einstein Archives 60-058

When you buy a piece of land to plant your cabbage and apples, you first have to drain it; that will kill all forms of animal and plant life that exist in that water. Later you would have to kill all the worms and cat-erpillars etc. that would eat your plants. If you must avoid all this killing on moral grounds, you will in the end have to kill yourself, all for the sake of leav-ing alive those creatures who have no such concep-tion of higher moral principles.

Ibid. Quoted in *Vegetarisches Universum*, December 1957

So I am living without fats, without meat, without fish, but am feeling quite well this way. It almost seems to me that man was not born to be a carnivore.

To Hans Muehsam, March 30, 1954. Einstein Archives 38-435. Einstein was probably not a vegetarian by choice, for he left behind no remarks that it was a moral issue for him. He had lifelong stomach problems that required him to watch his diet carefully. He may have avoided alcohol for the same reason.

Violence

Violence may at times have quickly cleared away an obstruction, but it has never proved itself to be creative.

> Comment on an article by J. Krutch, "Was Europe a Success?" (1934). Reprinted in *Einstein on Humanism*, 49. Einstein Archives 28-282

Wealth

One should not forget that wealth has its obligations.

> To Heinrich Zangger, March 26, 1920. *CPAE*, Vol. 9, Doc. 361

I am absolutely convinced that no amount of wealth can help humanity forward, even in the hands of the most dedicated worker in this cause. The example of great and pure personalities can lead us to noble deeds and views. Money only appeals to selfishness, and, without fail, it tempts its owner to abuse it. Can anyone imagine Moses, Jesus, or Gandhi with the moneybags of Carnegie?

> From "On Wealth," December 9, 1932, for *Die Bunte Woche*. Published in *Mein Weltbild* (1934), 10–11; reprinted in *Ideas and Opinions*, 12–13

The economists will have to revise their theories of value.

Upon being told that two of his handwritten manuscripts sold for $11.5 million at an auction for the war bonds effort. Recounted by historian Julian Boyd to Dorothy Pratt, February 11, 1944, Princeton University Archives; quoted in Sayen, *Einstein in America*, 150

All I want in my dining room is a pine table, a bench, and a few chairs.

Quoted in Maja Einstein's biography of her brother, in *CPAE*, Vol. 1; also quoted in Dukas and Hoffmann, *Albert Einstein, the Human Side*, 14

Wisdom

Wisdom is not a product of schooling but of the life-long attempt to acquire it.

To J. Dispentiere, March 24, 1954. Einstein Archives 59-495

Women

We men are deplorable, dependent creatures. But compared with these women, every one of us is king, for he stands more or less on his own two feet, not constantly waiting for something outside of himself to cling to. They, however, always wait for someone to come along who will use them as he sees fit. If this does not happen, they simply fall to pieces.

To Michele Besso, July 21, 1917, in a discussion about Einstein's wife, Mileva. *CPAE*, Vol. 8, Doc. 239

Very few women are creative. I would not send a daughter of mine to study physics. I'm glad my wife doesn't know any science. My first wife did.

> Quoted by Esther Salaman, who had been a young student in Berlin when Einstein was there, in the *Listener*, September 8, 1968; also quoted in Highfield and Carter, *The Private Lives*, 158

As in all other fields, in science the way should be made easy for women. Yet it must not be taken amiss if I regard the possible results with a certain amount of skepticism. I am referring to certain restrictive parts of a woman's constitution that were given her by Nature and which forbid us from applying the same standard of expectation to women as to men.

> 1920. Quoted by Moszkowski, *Conversations with Einstein*, 79

When women are in their homes, they are attached to their furnishings . . . they are always fussing with them. When I am with a woman on a trip, I am the only piece of furniture she has available, and she cannot refrain from circling around me all day and making some improvements on me.

> Quoted in Frank, *Einstein: His Life and Times*, 126. Einstein was prone to making wisecracks such as this one.

Work

Work is the only thing that gives substance to life.

> To son Hans Albert, January 4, 1937. Einstein Archives 75-926

It is really a puzzle what drives one to take one's work so devilishly seriously. For whom? For oneself? One soon departs this world, after all. For one's contemporaries? For posterity? *No*. It remains a puzzle.

> To artist Joseph Scharl, December 27, 1949. Einstein Archives 34-207

I am also convinced that one gains the purest joy from spiritual things only when they are not tied in with earning one's livelihood.

> To L. Manners, March 19, 1954. Einstein Archives 60-401

Youth

Truly novel ideas emerge only in one's youth. Later on one becomes more experienced, famous—and foolish.

> To Heinrich Zangger, December 6, 1917. *CPAE*, Vol. 8, Doc. 403

When I read your letters, I am very much reminded of my youth. In one's thoughts, one tends to set oneself against the world. One compares one's strengths with everything else, one alternates between despondency and self-assurance. One has the feeling that life is eternal and that everything one does and thinks is so important.

> To son Eduard, 1926. Einstein Archives 75-645

O, Youth: Do you know that yours is not the first generation to yearn for a life full of beauty and freedom? Do you know that all your ancestors have felt the same as you do—and fell victim to trouble and hatred? Do you know also that your fervent wishes can only find fulfillment if you succeed in attaining a love and an understanding of people, and animals, and plants, and stars, so that every joy becomes your joy and every pain your pain?

Written in I. Stern's autograph album in Caputh, Germany, 1932. Quoted in Dukas and Hoffmann, *Albert Einstein, the Human Side*, 129

Einstein's Verses: A Small Selection

I realize that much may be lost in translation, especially in poetry, and maybe it is a mistake to include in this volume some verses translated from Einstein's meticulous German into English. However, a number of people have asked me to do so. Though many of these translations, gleaned from a variety of sources, sound awkward, readers may get a sense of the extent—and sometimes the limitations—of Einstein's poetic talents. The rhythm and rhyme of most of the poems are similar to those of the German poet Wilhelm Busch of "Max und Moritz" fame, and most are infused with humor, often tongue-in-cheek. Unfortunately, not all of these translations have that effect. The verses, poems, and limericks in the Einstein Archives number in the hundreds. Most are untitled and were often sent to Einstein's friends in place of a letter, or added to a photograph or postcard. I've added the translator's name if it is known.

*Wherever I go and wherever I stay
There's always a picture of me on display.
On top of the desk, or out in the hall,
Tied round a neck, or hung on the wall.
Women and men, they play a strange game,
Asking, beseeching: "Please sign your name."
From the erudite fellow they brook not a quibble
But firmly insist on a piece of his scribble.

Sometimes, surrounded by all this good cheer,
I'm puzzled by some of the things that I hear,

And wonder, my mind for a moment not hazy,
If I and not they could really be crazy.

Dedicated to Cornelia Wolff, January 1920. In Dukas and
Hoffman, *Albert Einstein, the Human Side*, 73–74

*A little technology here and there
Can interest thinkers everywhere
And so I boldly think ahead:
The two of us will lay an egg.

In a letter to Rudolf Goldschmidt, November 1928, request-
ing a collaboration. Trans. by Aaron Wiener in Neffe, *Einstein*, 43. Einstein Archives 31-071

*Men and women of all ages
Leave your mark upon these pages.
But not with prose resembling that
Of people as they stop to chat;
Arrange your words in flowing lines
Like lofty poets' neat designs.
Begin, and push your fears away
Your writing will not go astray.

Preface to the guestbook at the house in Caputh, May 4,
1930. Trans. by Aaron Wiener in Neffe, *Einstein*, 306. Einstein Archives 31-067

*Long-branched and delicately strung,
Nothing will escape her gaze.

The welcome sight of her friends
Smiling, and still a weeping willow.

To one of Einstein's numerous lady friends, Ethel Michanowski, May 16, 1931. Einstein Archives 84-103

*Duty in mind, pipe in hand
That's how Captain Trauernicht stands.
Smiling wide, with eyes ablaze
Nothing can escape his gaze.
He looks out on ship and sea
His crew obeys him 1–2–3
Calmly, Trauernicht stands his ground
And takes in everything around.

Written ca. 1932. Trans. by Aaron Wiener in Neffe, *Einstein*, 28. According to Barbara Wolff of the Einstein Archives, "Trauernicht" ("Carefree" in the translation) was the captain's real name. Einstein Archives 31-099

*Whoever has let a daughter of Eve
Into his black heart
Feels sorry for the day that flies by quietly
When he can't catch a glimpse of her.

To Ethel Michanowski, 1932 or 1933. Einstein Archives 84-108

*A thousand letters in the mail
And every journal tells his tale
What's he to do when in this mood?
He sits and hopes for solitude.

Written in 1934. Trans. by Aaron Wiener in Neffe, *Einstein*, 16. Einstein Archives 31-161

*All my friends are hoaxing me
—Help me stop the family!
Reality's enough for me,
I've borne it long and faithfully.
And yet it would be nice to find
That I'd possessed the strength of mind
To lay eggs on the side—so long
As others didn't take it wrong.
                    A. Einstein, Stepfather

Written in 1936, to friend János Plesch, after Einstein heard a rumor that he had fathered an illegitimate child. Translation in Highfield and Carter, *Private Lives*, 93–94. Einstein Archives 31-178

*The postman brings me every day
Piles of mail, to my dismay.
Oh, why does no one ever reason
That he is one, while we are legion.

Written in 1938. Quoted by Peter Bucky in Neffe, *Einstein*, 373. Einstein Archives 31-215

*Even if one loves to play
One's little fiddle night and day
It's not right to broadcast it
Lest the list'ners scoff at it.
If you scratch with all your might—
Which is certainly your right—

Then bring down the windowpane
So the neighbors don't complain.

Poem to Emil Hilb, April 18, 1939 (my translation). Einstein
Archives 31-279

The Wisdom of Dialectical Materialism

*Through sweat and effort beyond compare
To arrive at a small grain of truth?
A fool is he who toils to find
What we simply ordain as the Party line.

And those who dare to express doubt
Will quickly find their skulls bashed in.
And thus we educate as never before
Bold spirits to embrace harmony.

Written in 1952. On the same page Einstein scrawled a sar-
castic aphorism, "Inscription for the Marx-Engels Institute:
In the realm of truth-seekers there is no human authority.
He who attempts to play the ruler there will run afoul of
the laughter of the gods"; see Rowe and Schulmann, *Einstein on Politics*, 457. The aphorism is translated a bit differently in the source used in the section "On Humankind" in
this volume. Einstein Archives 28-948

*How much I love that noble man!
More than I can tell with words.
I fear, however, he'll stand alone
With a halo all his own.

On the Jewish philosopher Baruch Spinoza. Einstein Archives 33-264

That little word "we" I mistrust and here's why:
No man of another can say he is I.
Behind all agreement lies something amiss
All seeming accord cloaks a lurking abyss.

Verse quoted in Dukas and Hoffmann, *Albert Einstein, the Human Side*, 100

*Whoever writes grim fairy tales
Will end up in our harshest jails.
But if he dares the truth to tell
We'll cast his soul down into hell.

Quoted by Peter Bucky in Neffe, *Einstein*, 285

*The Jews around me that I see
Little pleasure give to me.
When the others I then view
I am glad to be a Jew.

Einstein Archives 31-324. Trans. by Josef Eisinger. Quoted in Neffe, *Einstein*, 321

The following poems were written for Johanna Fantova, Einstein's last lady-friend, ca. 1947 to 1955. Translations are my own (with my apologies). The original German versions and different translations can be found in *Princeton University Library Chronicle* 65, no. 1 (Autumn 2003).

*It's Saturday. I sit alone
With notebook and a lamp aglow.

On the table lies my pipe,
Off to bed, it's late at night!

*It doesn't help to brood and fret
Hanne has gone out, she said.
True, the errands must be run
For her it certainly is no fun.

*I, most clumsy of them all,
Would have taken quite a fall.
If you hadn't helped right there,
I would be in great despair.

*What good is 2-8-4-2-J
If silent it resolves to stay?
Each day I toil at home alone,
One's like an orphan with no phone.

[Written on a day when the phones were out of order.]

Attributed to Einstein

Over the past fifteen years, a number of people have sent me the sources of quotations that appeared in this section in the previous editions. I thank them for this help, and I have inserted the quotations into the text in the appropriate sections. I'm still looking for the sources of many of the following quotations. Some sound genuine, some are apocryphal, and others are no doubt fakes, created by those who wanted to use Einstein's name to lend credibility to a cause or an idea. Hundreds can be found on the Internet, on calendars, and in little books containing undocumented quotations, but I include here only quotations sent to me by curious readers.

Misattributed to Einstein

International law exists only in textbooks on international law.

> Actually said by Ashley Montagu in his interview with Einstein. See Montagu's "Conversations with Einstein," *Science Digest*, July 1985

Education is that which remains, if one has forgotten everything he learned in school.

> Not originally said by Einstein, though he agreed with it. He quoted this passage by an anonymous "wit" in "On Education," in *Out of My Later Years*, 38. A similar statement was made by Alan Bennett in *Forty Years On*, quoted in *Oxford Dictionary of Humorous Quotations* (2001), and perhaps by others as well.

Two things inspire me to awe—the starry heavens above and the moral universe within.

> This is an inaccurate version of one of Kant's most famous statements: "Two things fill the mind with ever new and increasing wonder and awe, the more often and steadily we reflect upon them: the starry heavens above me and the moral law within me," in *The Critique of Practical Reason* (New York: Open Court, 1969), 162

The wireless telegraph is not difficult to understand. The ordinary telegraph is like a very long cat. You pull the tail in New York, and it meows in Los Angeles. The wireless is the same, only without the cat.

> According to Barbara Wolff of the Einstein Archives, this is an old Jewish joke and can be found in various compendia.

*Insanity is doing the same thing over and over again and expecting different results.

> By Rita Mae Brown, in *Sudden Death* (New York: Bantam, 1983), 68. Thanks to Barbara Wolff for the source.

We use only 10 percent of our brains.

> This myth is repeated frequently and is false. Several articles have been written about it and were sent to me by readers. See Michael Brand and Grace A. Reband, "Missing: 90% of the Human Brain," in www.chicagoflame.com, accessed 1/15/2008

As a young man, my fondest dream was to become a geographer. However, while working in the customs office, I thought deeply about the matter and concluded that it was far too difficult a subject. With some reluctance, I turned to physics as a substitute.

> Sent by a member of a geography department in South Dakota, who said this quotation is circulating around many geography departments. Einstein was obviously already a physicist by the time he worked in the Patent (not Customs) Office, so it was a bit late to be thinking of geography as a possible future career.

Possibly or Probably by Einstein

Everything should be made as simple as possible, but not simpler.

> This quotation prompts the most queries; it appeared in *Reader's Digest* in July 1977, with no documentation. Everyone seems to know it, yet no one can find its original source. There appear to be several variants of it, too, most commonly: "A *theory* should be made as simple as possible, but not simpler." Occam's Razor, also known as the "principle of parsimony," is a scientific and philosophic rule that says that entities should not be multiplied unnecessarily. This is interpreted as requiring that the simpler of competing theories be preferred to the more complex, or that explanations of unknown phenomena be sought first in terms of known quantities. It could be that someone misattributed Occam's Razor to Einstein, and that this saying is actually over six hundred years older than we think it is. (William of Ockham lived ca. 1285–1349.) Isaac Newton, too, was a fan of simplicity: "Nature is pleased with simplicity, and affects not the pomp of superfluous causes," he wrote. Most likely, the quotation is a paraphrase of some of Einstein's other statements about simplicity, many of which can be found in this book. Einstein's remark, if Einstein did say it, of course cannot be taken literally.

For international communication, the growth of international understanding, with the help of an international language, is not only necessary but self-evident. Esperanto is the best solution of the idea of an international language.

> This quotation is probably genuine, though I can't find its source. The Association Mondiale Anationale, an Esperanto organization, was founded in Prague in 1921. Einstein accepted the honorary presidency of the group in 1923 in Kassel, so it is likely that he did make this remark. It complements his ideas on world government as well. Esperanto was forbidden both by the Nazis and by Stalin, and its German proponents were sent to concentration camps.

No amount of experimentation can ever prove me right; a single experiment can prove me wrong.

> This may be a paraphrase of sentiments expressed in "Induction and Deduction," December 25, 1919, *CPAE*, Vol. 7, Doc. 28

The significant problems we face cannot be solved at the same level of thinking that was used when we created them. (An inelegant variant: The world we have created today as a result of our thinking thus far has problems which cannot be solved by thinking the way we thought when we created them.)

> Another common query. Perhaps these are paraphrases of his 1946 quotation, "A new type of thinking is essential if mankind is to survive and move toward higher levels" or "Past thinking and methods did not prevent world wars. Future thinking must" (see section "On Peace, War, the

Bomb, and the Military"). See Schulmann and Rowe, *Einstein on Politics*, 383

At such moments one imagines that one stands on some spot of a small planet gazing in amazement at the cold yet profoundly moving beauty of the eternal, the unfathomable. Life and death flow into one, and there is neither evolution nor eternity, only Being.

Quoted by psychologist Deepak Chopra, *Ageless Body, Timeless Mind* (1993), 280, with no source

Nothing will benefit human health and increase the chances for survival on Earth as much as the evolution of a vegetarian diet.

This might be a modified variant of the similar statement in Einstein's letter to Hermann Huth (see "Vegetarianism" quotations).

*The intuitive mind is a sacred gift, and the rational mind is a faithful servant. We have created a society that honors the servant and has forgotten the gift.

The substance of our knowledge resides in the detailed terminology of a field.

In science the work of the individual is so bound up with that of his scientific predecessors and contemporaries that it appears almost as an impersonal product of his generation.

In any conflict between humanity and technology, humanity will win.

*The difference between genius and stupidity is that genius has its limits. (Variant: Two things are infinite: the universe and human stupidity; and I'm not sure about the universe.)

> Similar to Gustave Flaubert's "Human stupidity is infinite," written to Guy de Maupassant, February 19, 1880. Thanks to Cécile Caccamo for this source.

*The only way to make money at a roulette wheel is to steal it when the dealer isn't looking.

> Sent by an Australian reader, who believed he read somewhere that Einstein had visited a casino and expressed interest in the mechanics of the roulette wheel. I was not able to confirm the story.

*If I could remember the names of all those particles, I'd be a botanist.

> Said to be from "Science, Philosophy and Religion," but I could not find it in the reprinted speech in *Ideas and Opinions.*

If you think intelligence is dangerous, try ignorance.

> Similar to Derek Bok's "If you think education is expensive, try ignorance." See *Random House Webster's Quotationary* (1998).

Compounded interest is more complicated than relativity theory. (Variants: The most powerful force in

the universe is compound interest. The most significant invention of the nineteenth century is compounded interest.)

> Quoted by various economists, including Burton Malkiel
> in the *Princeton Spectator*, May 1997, and on several finan-
> cial Web sites on the Internet. (Thanks to Steven Feldman
> for showing me the variant.)

The most difficult thing to understand is the income
tax. (Variants: The hardest thing to understand in
the world is the income tax. The theory of relativity
is easy; the IRS code is difficult. Preparing a tax return is more complicated than relativity theory.)

> Quoted in the *Macmillan Book of Business and Economics
> Quotes* by M. Jackson (1984). No source given. This quota-
> tion no doubt comes from the cartoon introducing this sec-
> tion: "Try to make out my theory and your income tax
> work will look simple." In other words, this is the cartoon-
> ist's sentiment, not Einstein's.

If the bee becomes extinct, mankind will have only
four years to live: no bees, no pollination, no plants,
no animals, no humans.

> According to the Web site, www.snopes.com, this quota-
> tion started to pop up around 1994 during a protest by bee-
> keepers in Brussels in a pamphlet distributed by the Na-
> tional Union of French Apiculture. More likely, they
> distorted Einstein's letter of December 12, 1951 to school-
> children (see section "On and to Children").

Astrology is a science in itself and contains an illuminating body of knowledge. It taught me many

things and I am greatly indebted to it. Geophysical evidence reveals the power of the stars and the planets in relation to the terrestrial. This is why astrology is like a life-giving elixir to mankind.

> An excellent example of a quotation someone made up and attributed to Einstein in order to give credibility to an idea or cause. See Denis Hamel, "The End of the Einstein-Astrologer-Supporter Hoax," *Skeptical Inquirer* 31, no. 6 (November/December 2007), 39–43, for a thorough refutation of this myth.

Einstein's "Rules of Work":

1. Out of clutter, find simplicity.
2. From discord, find harmony.
3. In the middle of difficulty lies opportunity.

> The first "rule" is probably a paraphrase of Einstein's many quotations about the value of simplicity. I traced the second rule to Horace, the Roman poet and satirist, who had it as *Concordia discors* (harmony in discord) in his *Epistles* I, xii.19. And the third rule has probably been in general use for ages.

*If at first the idea is not absurd, then there's no hope for it.

*I speak to everyone in the same way, whether he is the garbage man or the president of the university.

*The truth of a theory is in your mind, not your eyes.

*The world needs heroes, and it's better they be harmless men like me than villains like Hitler.

* "I am no Einstein," said Einstein modestly.

*Common sense is the collection of prejudices acquired by age eighteen.

*When asked how he was different from the average person, Einstein replied: "When asked to look for a needle in a haystack, the average person would search until he found the needle; whereas I would look for all needles."

*Setting an example is not the main means of influencing another, it is the *only* means.

*Compounded interest is the eighth wonder of the world. (Variants: Compounding interest is the greatest mathematical discovery of all time. Compounding is mankind's greatest invention because it allows for the reliable, systematic accumulation of wealth. The most powerful force in the universe is compounded interest.)

*It's not that I'm so smart, it's just that I stay with problems longer.

*Logic will get you from A to B. Imagination will get you everywhere.

*The only reason there is time is so that everything doesn't happen at once.

*To have a better life we must keep choosing how we are living.

*Nothing happens unless something is moved.

*You do not really understand something unless you can explain it to your grandmother.

*Men marry women with the hope they will never change. Women marry men with the hope they will change. Invariably they are both disappointed.

*Any man who can drive safely while kissing a pretty girl is simply not giving the kiss the attention it deserves.

*Bureaucracy is the death of all sound work.

*Resolution 1: I will live for God. Resolution 2: If no one else does, I still will.

*Life is a mystery to be lived, not a problem to be solved.

The only source of knowledge is experience.
Playing is the highest form of research.
Individuality is an illusion created by skin.
Not everything that counts can be counted, and not everything that can be counted counts.
Knowledge is experience. Anything else is information.
There is no hope for an idea that at first does not seem insane.
The religion of the future will be a cosmic religion. . . . Buddhism answers this description.

The probability of life originating by accident is comparable to the probability of the unabridged dictionary resulting from an explosion in a print shop. (Variant: The idea that this universe in all its millionfold order and precision is the result of blind chance is as credible as the idea that if a print shop blew up all the type would fall down again in the finished and faultless form of the dictionary.)

If the facts don't fit the theory, change the facts.

There are only two ways to live your life. One is as though nothing is a miracle. The other is as though everything is a miracle.

As the circle of light increases, so does the circumference of darkness.

The practice of science as a whole finds truths and leads to a correct understanding of the universe, but science in actual practice is riddled with mistakes and the residue of human frailties.

The lowest level of awareness is "I know." Then, "I don't know," "I know I don't know," "I don't know that I don't know."

The levels of intelligence are "Smart, intelligent, brilliant, genius, *simple*."

Death means that one can no longer listen to Mozart.

*Einstein's Riddle:

There are five houses, painted in five different colors. A person with a different nationality lives in each house. Each of the five owners drinks a certain type of beverage, plays a certain sport, and keeps a specific pet. None of the owners has the same pet, plays the same sport, or drinks the same beverage. *Who owns the fish?* Here are the facts:

1. The Briton lives in the red house.
2. The Swede keeps dogs as pets.
3. The Dane drinks tea.
4. The green house is on the left of the white house.
5. The owner of the green house drinks coffee.
6. The person who plays football raises birds.
7. The owner of the yellow house plays baseball.
8. The man living in the center house drinks milk.
9. The Norwegian lives in the first house.
10. The man who plays volleyball lives next to the one who keeps cats.
11. The man who keeps the horse lives next to the man who plays baseball.
12. The owner who plays tennis drinks beer.
13. The German plays hockey.
14. The Norwegian lives next to the blue house.
15. The man who plays volleyball has a neighbor who drinks water.

To solve the problem, make grids: one column for each of the five houses, and five rows for nationality, house color, type of beverage, type of sport, and kind of pet. *SOLUTION (don't peek!)*: The fish is owned by the German who lives in house 4 (green), drinks coffee, and plays hockey.

> Supposedly, but not actually, devised by Einstein as a child. Allegedly only 2 percent of the population can solve it, but the solution appears to be attainable by those who are patient and persistent. I include it here for fun and because it has been attributed to Einstein. Many versions can be found on the Internet. This version is taken mostly from the book of the same name by Jeremy Stangroom.

Others on Einstein

With the passing of Dr. Albert Einstein, the world has lost its greatest scientific mind, the human race one of its most ethical and inspiring personalities, the Jewish people one of its most loyal sons.

Samuel Belkin, president of Yeshiva University, 1955. Quoted in Cahn, *Einstein*, 121

Tell me what to do if he says yes. I had to offer the post to him because it is impossible not to. But if he accepts, we are in for trouble.

David Ben Gurion to Yitzak Navon, after Israeli ambassador to the United States, Abba Eban, was instructed to offer the presidency of Israel to Einstein in November 1952. Quoted in Holton and Elkana, *Albert Einstein: Historical and Cultural Perspectives*, 295

For a long time, I have been thinking of how many . . . ties . . . bind the two of us together. I owe to you my wife, and along with her, my son and grandson; I owe to you my job, and with that, the tranquility of a sanctuary . . . as well as the financial security for hard times. I owe to you the scientific synthesis that I would never have acquired without such a friendship. . . . On my side, I was your audience in 1904 and 1905. In helping you edit your communications on quanta I deprived you of a part of your glory.

Michele Besso to Einstein, January 17, 1928. Quoted in Jeremy Bernstein, "A Critic at Large," *New Yorker*, February 27, 1989. Einstein Archives 7-101. Einstein had introduced

Besso to his future wife and recommended him for the job
he held in the Swiss patent office in Bern for many years.

When something struck him as funny, his eyes twin-
kled merrily and he laughed with his whole being. . . .
He was ready for humor.

Algernon Black, 1940. Einstein Archives 54-834

*Einstein, as you know, is no Zionist, and I ask you
not to try to make him a Zionist or to try to attach
him to our organization. . . . Einstein, who leans toward socialism, feels very involved with the cause
of Jewish labor and Jewish workers. [He] . . . often
says things out of naiveté which are unwelcome
by us.

Kurt Blumenfeld, in *The War about Zionism* (Stuttgart:
Deutsche Verlagsanstalt, 1976), 65–66. Cited in Jerome, *Einstein on Israel and Zionism,* 25

Through Albert Einstein's work the horizon of mankind has been immeasurably widened, at the same
time as our world picture has attained a unity and
harmony never dreamed of before. The background
for such an achievement was created by preceding
generations of the world community of scientists,
and its full consequences will only be revealed to
coming generations.

Physicist Niels Bohr. Quoted in Einstein's obituary in the
New York Times, April 19, 1955

The memory of his noble personality will always remain a fresh source of inspiration and strength to those of us who were happy enough to become personally acquainted with him.

Niels Bohr after Einstein's death, 1955. Quoted in Cahn, *Einstein*, 122

*Now his living voice is silent, but those who heard it will hear it until the end of their days.

Hedi Born, wife of physicist Max Born and a close friend, in the Borns' *Der Luxus des Gewissens* (*Luxury of a Conscience*) (Munich: Nymphenburger, 1969). Quoted by their son Gustav Born in *The Born-Einstein Letters*, vi

Einstein would be one of the greatest theoretical physicists of all time even if he had not written a single line on relativity.

Physicist and close friend Max Born. Quoted in Hoffmann, *Albert Einstein: Creator and Rebel*, 7

He knew, as did Socrates, that we know nothing.

Max Born, after Einstein's death. Quoted in Clark, *Einstein*, 415

*He saw in the quantum mechanics of today a useful intermediate stage between the traditional classical physics and a still completely unknown "physics of the future" based on general relativity in which . . . the traditional concepts of physical reality

and determinism come into their own again. Thus he regarded statistical quantum mechanics to be not wrong but "incomplete."

Max Born, in *Born-Einstein Letters*, 199

*Time and time again [Einstein] filled me with amazement, and indeed enthusiasm, as I watched the ease with which he would, in discussion, experimentally change his point of view, at times tentatively adopting the opposite view and viewing the whole problem from a new and totally changed angle.

Writer and editor Max Brod, whom Einstein befriended in Prague. In Brod, *Streitbares Leben* (Munich, 1969). See also Fölsing, *Einstein*, 283

He always took his celebrity with humor and laughed at himself.

Family friend Thomas Bucky, shown in A&E Television's Einstein biography, VPI International, 1991

*The bright boys here all study Math,
And Albie Einstein points the path.
Although he seldom takes the air,
We wish to God he'd cut his hair.

From "The Faculty Song" of the Princeton songbook *Carmina Princetonia* (an annual event through 1968), precise date unknown. Thanks to Trevor Lipscombe for this submission.

Certainly he was a great savant, but beyond that he was also a pillar of human conscience at a moment when so many civilizing values seemed to be in the balance.

Pablo Casals to Carl Seelig. Quoted in French, *Einstein*, 43. Einstein Archives 34-350

Among twentieth-century men, he blends to an extraordinary degree those highly distilled powers of intellect, intuition, and imagination which are rarely combined in one mind, but which, when they do occur together, men call genius. It was all but inevitable that this genius should appear in the field of science, for twentieth-century civilization is first and foremost technological.

Whittaker Chambers, in a *Time* magazine cover story, July 1, 1946

They cheer me because they all understand me, and they cheer you because no one understands you.

Actor Charlie Chaplin, after the premiere of *City Lights* in Los Angeles in January 1931, to which Chaplin had invited Einstein. See Fölsing, *Albert Einstein*, 457

His short skull seems unusually broad. His complexion is matte light brown. Above his large sensuous mouth is a thin black mustache. The nose is slightly aquiline. His striking brown eyes radiate deeply and softly. His voice is attractive, like the vibrant note of a cello.

> Einstein's student Louis Chavan. Quoted in Max Flückiger,
> *Albert Einstein in Bern* (1972), 11–12

*[Einstein's death] is a loss of righteous mankind. His contributions to science are epoch-making. He loved science and human beings. . . . He strove untiringly for the cause of peace, democracy, and freedom. With the greatest grief, the Chinese people mourn this outstanding scientist and great soldier for the cause of human peace.

> Peiyuan Chou, president of the Chinese Physical Society,
> *People's Daily* (Beijing), April 21, 1955. Quoted in Hu, *China
> and Albert Einstein*, 144

*Albert Einstein Held Me in His Arms
by David Clewell

although my parents didn't know it at the time.
And if I knew anything, even on some vaguely
　　molecular level,
I surely wasn't talking. No one was the wiser,
　　except
for Einstein, of course, taken with my small
　　charms.
He was crazy about how I couldn't stop smiling,
drooling in my carriage on a Sunday afternoon in
　　Princeton—the town my mother loved just
　　driving to and getting out and

losing herself in, absolutely smitten. And my
   pedestrian father
was crazy about my mother, so even if that meant
another goddamn trip to Highfalutinsville, New
   Jersey,
he'd be there without fail, forever along for the
   ride.
The way I finally heard it, Einstein was on his
   knees
in a sweatshirt, rumpled chinos, and sneakers,
   pulling weeds—
Merely being himself, my father would say later,
   utterly impressed.
Einstein had that down to a science at 112 Mercer,
   the unassuming
white frame house where he cultivated flowers,
   where he played violin
precisely in sync with his favorite recordings late
   into the night.
Where he famously met with Bertrand Russell,
   Kurt Gödel, and
Wolfgang Pauli for philosophic forays into the
   schnapps, then inevitably higher mathematics.
But on that one historic Sunday in the spring of my
   first year,
Einstein himself welcomed the unrenowned likes
   of my mother and father.
This twentieth-century giant picked me up with
   some easy peekaboo

small talk in the last of the afternoon's fading light
　　until, eventually, genius
or no genius, I couldn't take it anymore, and I
　　made a tiny grab
for his wildly theoretical hair. And that was pretty
　　much the end
of our ad hoc civilization that flourished for ten
　　Princeton minutes.
When Einstein died only six weeks later, every
　　newspaper ran his picture, and all at once my
　　father couldn't believe it: *Wasn't that the gardener
　　who couldn't get enough*
*of the baby? It says right here he's Einstein, the guy who
　　revolutionized our thinking about time and space!*
And what was that supposed to mean to him,
　　exactly? My father
wasn't Einstein, but he'd thought about them
　　plenty, too,
deciding in his lifetime he wasn't about to get
　　enough of either one.
For years my parents never said a word about that
　　day, as if
to remember it out loud would have been
　　somehow unseemly—
a kind of bragging they never much went in for—
　　rather than a celebration
of wonderful dumb-luck Sunday driving, like
　　every happy accident
in the history of science or in those classic, unlikely
　　stories

we can't help going back to for their mythic staying
     power. So now let me
put it this way: Albert Einstein held me in his arms
     before he died.
Sooner or later we're all trying to explain our
     particle selves
in light of our own cockeyed theories of relativity.
Someone in my family—my mother or my father,
     maybe me—
had to embellish at least some of the truth that
     comes, finally,
here at the end:
my mother's horrified
that I've yanked poor Einstein's hair, and she
     resigns herself,
sighing, *It's time to go.* To prove there are no hard
     feelings,
he says something Einsteinian, like *Yes, but what is
     time?*—
which my father misunderstands as a question he
     can actually answer
at that very minute, so he says, *5:00.* And before I
     know it,
because I am far too young to realize much of
     anything,
everyone's in a sudden hurry back into their
     uncertain futures,
as if this whole thing never quite honestly
     happened, and in no time

it's fifty years later, and I'm the one still alive, all
    that's left
of the story, telling myself: Yes, it did. No, it didn't.
    No, it did.

> The poem first appeared in the Summer 2006 issue of *The
> Georgia Review*. Thanks to Judith May for sending it to me.
> Reprinted with permission of David Clewell.

His eyes watered almost continually; even in moments of laughter he would wipe away a tear. . . .
The contrast between his soft speech and ringing
laughter was enormous. . . . Every time he made a
point that he liked, or heard something that appealed to him, he would burst into a booming laughter that would echo from the walls.

> From I. Bernard Cohen, "Einstein's Last Interview," April
> 1955. Published in *Scientific American* 193, no. 1 (July 1955),
> 69–73; reprinted in Robinson, *Einstein*, 212–225

Einstein is great because he has shown us our world
in truer perspective and has helped us to understand a little more clearly how we are related to the
universe around us.

> Physics Nobelist Arthur Compton. Quoted in Cahn, *Ein-
> stein*, 88

I was able to appreciate the clarity of his mind, the
breadth of his information, and the profundity of

his knowledge. . . . One has every right to build the greatest hopes on him and to see in him one of the leading theoreticians of the future.

> Marie Curie, in a letter to Pierre Weiss, November 17, 1911. Quoted in Hoffmann, *Albert Einstein: Creator and Rebel*, 98–99

One cannot contemplate without astonishment and admiration work at once so profound and so powerfully original achieved in a few years.

> Prince Louis de Broglie, permanent secretary of the French Academy of Science until his death. Quoted in Cahn, *Einstein*, 121

Your mere presence here undermines the class's respect for me.

> Said to Einstein by his seventh-grade teacher, Dr. Joseph Degenhart, who also predicted that he "would never get anywhere in life." In draft of a letter to Philip Frank, 1940. Einstein Archives 71-191, where Einstein also states that he wanted to leave the school to join his parents in Italy. See also *CPAE*, Vol. 1, lxiii, for more on this topic.

The contributions which Dr. Einstein made to man's understanding of nature are beyond assessment in our day. Only future generations will be competent to grasp their full significance.

> Harold Dodds, president of Princeton University, 1955. Quoted in Cahn, *Einstein*, 122

The professor never wears socks. Even when he was invited by Mr. Roosevelt to the White House he didn't wear socks.

> Einstein's secretary, Helen Dukas. Related by Philippe Halsman in French, *Einstein: A Centenary Volume*, 27

Einstein was prone to talk about God so often that I was led to suspect he was a disguised theologian.

> Writer Friedrich Dürrenmatt, in *Albert Einstein: Ein Vortrag*, 12

After a careful study of the plates I am prepared to say that there can be no doubt that they confirm Einstein's prediction. A very definite result has been obtained that light is deflected in accordance with Einstein's law of gravitation.

> British Astronomer-Royal, Sir Frank Dyson, after the Eddington expedition in May 1919 had confirmed Einstein's general theory of relativity. *Observatory* 32 (1919), 391

Dr. Einstein the scientist and Einstein the Jew represent a perfect harmony. These considerations, added to his deep emotion at the Jewish disaster in Europe . . . , explain the ardent zeal with which he advocated and sustained Israel's national revival.

> Abba Eban, Israel's ambassador to the United States in the 1950s. Quoted in Cahn, *Einstein*, 92

Here we have nothing but people who love *you* and not just your cerebral cortex.

Close friend Paul Ehrenfest in a letter to Einstein, September 8, 1919

*[Einstein is] one of the wonders of Nature . . . [and] a marvelous interweaving of simplicity and subtlety, of strength and tenderness, of honesty and humor, of profundity and serenity.

Quoted by Martin Klein in *The Lesson of Quantum Theory* (North Holland, 1986), 329

*Intellecual women did not attract him. Out of pity he was attracted to women who did manual work.

Elsa Einstein, Albert's second wife; said to the attractive Vera Weizmann, wife of Chaim Weizmann. Quoted in *The New Palestine*, April 1, 1921, 1. In fact, Einstein had affairs with several intelligent women.

God has put so much into him that is beautiful, and I find him wonderful, even though life at his side is debilitating and difficult in every respect.

Elsa Einstein, in a letter to Hermann Struck and wife, 1929. Quoted in Fölsing, *Albert Einstein*, 429

It is not ideal to be the wife of a genius. Your life does not belong to you. It seems to belong to everyone else. Nearly every minute of the day I give to my husband, and that means to the public.

Elsa Einstein. Quoted in her obituary in the *New York Times*, December 22, 1936, two days after her death

Oh, my husband does that on the back of an old envelope!

> Elsa Einstein, after a host at Mount Wilson Observatory
> in California explained to her that the giant telescope is
> used to find out the shape of the universe. Reported by
> Bennett Cerf in *Try and Stop Me* (New York: Simon &
> Schuster, 1944). Also in *The Folio Book of Humorous Anecdotes*
> (2005)

Probably the only project he ever gave up on was me. He tried to give me advice, but he soon discovered that I was too stubborn and that he was just wasting his time.

> Hans Albert Einstein in the *New York Times*, July 27, 1973.
> Quoted in Pais, *Einstein Lived Here*, 199

He was very fond of nature. He did not care for large, impressive mountains, but he liked surroundings that were gentle and colorful and gave one lightness of spirit.

> Hans Albert Einstein. Quoted in an interview with Bernard
> Mayes, in Whitrow, *Einstein*, 21

He often told me that one of the most important things in his life was music. Whenever he felt he had come to the end of the road or into a difficult situation in his work, he would take refuge in music and that would usually resolve all his difficulties.

> Ibid.

His work habits were rather strange. . . . Even when there was a lot of noise, he could lie down on the sofa, pick up a pen and paper, precariously balance an inkwell on the backrest, and engross himself in a problem so much so that the background noise stimulated rather than disturbed him.

Maja Einstein. See *CPAE*, Vol. 1, lxiv

*Did you know that I was in the same hospital as Albert? I was allowed to see him twice more and talk to him for a few hours. . . . I did not recognize him at first—he was so changed by the pain and blood deficiency. But his personality was the same as ever. He was . . . completely in command of himself with regard to his condition; he spoke with a profound serenity—even with a touch of humor—about the doctors and awaited his end as an imminent natural phenomenon. As fearless as he had been all his life, so he faced death humbly and quietly.

Margot Einstein in a letter to Hedi Born, after April 1955. See Born, *Born-Einstein Letters*, 229

When one was with him on the sailboat, you felt him as an element. He had something so natural and strong in him because he was himself a piece of nature. . . . He sailed like Odysseus.

Margot Einstein, May 4, 1978, in an interview with Jamie Sayen, in Sayen, *Einstein in America*, 132

No other man contributed so much to the vast expansion of twentieth-century knowledge.

Statement by President Dwight D. Eisenhower upon Einstein's death. Quoted in Einstein's obituary in the *New York Times*, April 19, 1955

In view of his radical background, this office would not recommend the employment of Dr. Einstein, on matters of a secret nature, without a very careful investigation, as it seems unlikely that a man of his background could, in such a short time, become a loyal American citizen.

Recommendation by the Federal Bureau of Investigation (FBI), which had never been informed of Einstein's letter to President Roosevelt warning him about the possibility of the Germans' building a bomb. Quoted by Richard Schwartz in *Isis* 80 [1989], 281–284. See also Rowe and Schulmann, *Einstein on Politics*, 59

*In the case of someone like Einstein we cannot help but feel that there is indeed an inner and necessary connection between the extraordinary theoretical simplicity of his work and the personal simplicity of the man himself. We feel that only someone himself so simple could have conceived such ideas.

Henry LeRoy Finch of Sarah Lawrence College, June 1970. In Introduction to Moszkowski, *Conversations with Einstein*, xxiii

*He was as interested in physiques as in physics.

Heard by a viewer on Fox News, ca. July 2006

Einstein's conversation was often a combination of inoffensive jokes and penetrating ridicule so that some people could not decide whether to laugh or to feel hurt. . . . Such an attitude often appeared to be an incisive criticism, and sometimes even created an impression of cynicism.

Philipp Frank in Frank, *Einstein: His Life and Times*, 77

He, who had always had something of a bohemian in him, began to lead a middle-class life . . . in a household such as was typical of a well-to-do Berlin family. . . . When one entered . . . one found that Einstein still remained a "foreigner" in such a surrounding—a bohemian guest in a middle-class home.

Ibid., 124

He is cheerful, assured, and courteous, understands as much of psychology as I do of physics, and so we had a pleasant chat.

Sigmund Freud, 1926, on a visit to Berlin, where he met Einstein. In a letter to S. Ferenczi, January 2, 1927, in *The Collected Papers of Sigmund Freud*, ed. Ernest Jones (out of print)

[I have finished writing] the tedious and sterile so-called discussion with Einstein.

Sigmund Freud to Max Eitingon, September 8, 1932, on Freud and Einstein's correspondence published by the League of Nations in 1933 as *Why War?* In ibid., 175

Of course I always knew that you admired me only "out of politeness," and that you are convinced of very few of my assertions. . . . I hope that by the time you have reached my age you will have become a disciple of mine.

> From Sigmund Freud to Einstein, May 3, 1936, replying to Einstein's letter congratulating Freud on his eightieth birthday. In *Letters of Sigmund Freud*, ed. Ernst L. Freud (New York: Basic Books, 1960). Einstein Archives 32-567

Einstein's fiddling was like that of a lumberjack sawing a log.

> Comment by professional violinist Walter Friedrich. Quoted in Herneck, *Einstein privat* (Berlin, 1978), 129

Of course, the old man agrees with almost anything nowadays.

> Cosmologist George Gamow. Written on the bottom of a letter from Einstein of August 4, 1948, in which Einstein states that one of Gamow's ideas is probably correct. In Frederick Reines, ed., *Cosmology, Fusion and Other Matters* (Boulder: University Press of Colorado, 1972), 310

Mankind has lost its finest son, whose mind reached out to the ends of the universe but whose heart overflowed with concern for the peace of the world and the well-being, not of humanity as an abstraction, but of ordinary men and women everywhere.

> Israel Goldstein, president of the American Jewish Congress. Quoted in Cahn, *Einstein*, 122

I think it was his sense of reverence.

> Dean Ernest Gordon of Princeton University's chapel, when asked how he would explain Einstein's combination of great intellect with apparent simplicity. Quoted by Richards, "Reminiscences," in *Einstein as I Knew Him*

*Your visit [to England] has had more tangible results in improving here the relations between our two countries than any other single event. Your name is a power in our country.

> Lord Richard B. S. Haldane, June 26, 1921. *CPAE*, Vol. 12, Doc. 159

A man distinguished by his desire, if possible, to efface himself and yet impelled by the unmistakable power of genius which would not allow the individual of whom it had taken possession to rest for one moment.

> Lord Haldane. Quoted in *The Times* (London), June 14, 1921

Einstein has called forth a greater revolution in thought than even Copernicus, Galileo, or Newton himself.

> Lord Haldane. Quoted in Cahn, *Einstein*, 10

Someone told me that each equation I included in a book would halve its sales. I therefore resolved not to have any equations at all. In the end, however, I *did* put in one equation, Einstein's famous equation,

$E = mc^2$. I hope that will not scare off half my potential readers.

Stephen Hawking in Hawking, *A Brief History of Time* (London: Bantam, 1988), vi

Quark, Strangeness and Charm

Einstein was not a handsome fellow
Nobody ever called him Al
He had a long moustache to pull on, it was yellow
I don't believe he ever had a girl
One thing he missed out in his theory
Of time and space and relativity
Is something that makes it very clear
He was never gonna score like you and me
He didn't know about
Quark, Strangeness and Charm
Quark, Strangeness and Charm
Quark, Strangeness and Charm

I had a dangerous liaison
To have been found out would've been a disgrace
We had to rendezvous some days on
the corner of an undiscovered place
We got sick of chat chat chatter
And the look upon everybody's face
But all that doesn't not anti-matter now
We've found ourselves a black hole out in space
And we're talking about
Quark, Strangeness and Charm

Quark, Strangeness and Charm
Quark, Strangeness and Charm

Copernicus had those Renaissance ladies
Crazy about his telescope
And Galileo had a name that made his
Reputation higher than his hopes
Did none of those astronomers discover
While they were staring out into the dark
That what a lady looks for in her lover
Is Charm, Strangeness and Quark
And we're talking about
Quark, Strangeness and Charm
Quark, Strangeness and Charm
Quark, Strangeness and Charm

> Lyrics from the 1977 song by the British rock group
> Hawkwind. Thanks to Trevor Lipscombe for sending it
> to me. © 1977 Anglo Rock, Inc. Reprinted with permission.

Not yet hanged.

> The Hitler regime's caption under the photo of Einstein in
> its official book of photographs of its "enemies of the
> state," 1933. A bounty of 20,000 marks had been put on his
> head. See Sayen, *Einstein in America,* 17

The essence of Einstein's profundity lay in his simplicity; and the essence of his science lay in his
artistry—his phenomenal sense of beauty.

> Banesh Hoffmann in Hoffmann, *Albert Einstein: Creator and
> Rebel,* 3

When it became clear that [we could not solve a problem], Einstein would stand up quietly and say, in his quaint English, "I vill a little t'ink." So saying he would pace up and down or walk around in circles, all the time twirling a lock of his long, graying hair around his finger.

Banesh Hoffmann. Recollection quoted in Whitrow, *Einstein*, 75

He was one of the greatest scientists the world has ever known, yet if I had to convey the essence of Albert Einstein in a single word, I would choose *simplicity*.

Banesh Hoffman, opening words in "My Friend, Albert Einstein," *Reader's Digest*, January 1968

The "Great Relative."

Name given to Einstein by the Hopi Indians on his visit to the Grand Canyon of the United States, February 28, 1931, on his return trip by train from California to New York. Recounted in A&E Television's Einstein biography, VPI International, 1991. Also reported by Fölsing, *Albert Einstein*, 640

As Time Goes By

This day and age we're living in
Gives cause for apprehension
With speed and new invention
And things like fourth dimension.
Yet we get a trifle weary

With Mr. Einstein's theory.
So we must get down to earth at times
Relax relieve the tension.
And no matter what the progress
Or what may yet be proved
The simple facts of life are such
They cannot be removed.
You must remember this
A kiss is just a kiss, a sigh is just a sigh.
The fundamental things apply
As time goes by.

From the song "As Time Goes By," lyrics and music by Herman Hupfeld, made famous by the long-running British television series of that title, starring Dame Judi Dench, and the film *Casablanca* (1942). The first three verses, which pertain to Einstein, have been virtually unknown. (© 1931 [Renewed] Warner Bros. Inc. All rights administered by WB Music Corp. All rights reserved. Used by permission of Alfred Publishing Co., Inc., Los Angeles.)

Einstein gave his wife the greatest care and sympathy. But in this atmosphere of approaching death, Einstein remained serene and worked constantly.

Leopold Infeld, on Einstein's coping with wife Elsa's terminal illness of heart and kidney disease. In Infeld, *The Quest*, 282

The greatness of Einstein lies in his tremendous imagination, in the unbelievable obstinacy with which he pursues his problems.

Ibid., 208

One of my colleagues in Princeton asked me: "If Einstein dislikes his fame and would like to increase his privacy, why does he . . . wear his hair long, a funny leather jacket, no socks, no suspenders, no ties?" The answer is simple. The idea is to restrict his needs and, by this restriction, increase his freedom. We are slaves of millions of things. . . . Einstein tried to reduce them to the absolute minimum. Long hair minimized the need for the barber. Socks can be done without. One leather jacket solves the coat problems for many years.

Ibid., 293

For Einstein, life is an interesting spectacle that he views with only slight interest, never torn by the tragic emotions of love or hatred. . . . The great intensity of Einstein's thought is directed outside toward the world of phenomena.

Infeld, *Einstein*, 123

If Einstein were to enter your room at a party and he were introduced to you as a "Mr. Eisenstein" of whom you knew nothing, you would still be fascinated by the brilliance of his eyes, by his shyness and gentleness, by his delightful sense of humor, by the fact that he can twist platitudes into wisdom. . . . You feel that before you is a man who thinks for himself. . . . He believes what you tell him because he is kind, because he wishes to be kind, and because it is much easier to believe than to disbelieve.

Ibid., 128

*Einstein's English was quite simple and consisted of about three hundred words, which he pronounced very weirdly.

In Leopold Infeld, *Leben mit Einstein* (Vienna: Europa Verlag, 1969), 73. Translated in Neffe, *Einstein*, 35

*In matters of logic and contemplation, Einstein understood everyone quite well, but he had a much harder time grasping emotional issues. It was difficult for him to imagine impulses and feelings unrelated to his own life.

Ibid., 54. Translated in Neffe, *Einstein*, 372

Is $E = mc^2$ a sexed equation? . . . Perhaps it is. Let us make the hypothesis that it is insofar as it privileges the speed of light over other speeds that are vitally necessary to us. What seems to me to indicate the possible sexed nature of the equation is not directly its uses by nuclear weapons, rather it is having privileged what goes the fastest.

From an article by feminist Luce Irigaray, "Is the Subject of Science Sexed?" *Hypatia: A Journal of Feminist Philosophy* 2, no. 3 (1987), 65–87. Cited in Francis Wheen, *How Mumbo-Jumbo Conquered the World: A Short History of Modern Delusions* (London: Fourth Estate, 2004), 88

Einstein's attention was far from the macaroni we were eating.

> Russian physicist A. F. Joffe, after discussions of their mutual work during dinnertime at Einstein's home in Berlin. Recalled in *Die Wahrheit* (Berlin), March 15–16, 1969

Einstein had very few hobbies. One was puzzles, and he got the most amazing ones from all over the world. . . . I brought him the famous Chinese Cross, one of the most complicated puzzles to put together. He solved it in three minutes.

> Friend Alice Kahler, quoted in the *Princeton Recollector* (1985), 7. See Roboz Einstein, *Hans Albert Einstein*, 38

With all his phenomenal intellect, he is still a naïve and altogether spontaneous human being.

> Friend Erich Kahler, 1954. Einstein Archives 38-279

I can still see you . . . your kind face still sparkling with delight! This [childlike cheerfulness] seems to me a fine symbol of how well you will exert a lasting influence on scientific life here.

> From Heike Kamerlingh Onnes, February 8, 1920. *CPAE*, Vol. 9, Doc. 304

You'd better watch out, you'd better take care, Albert says E equals $m\ c$ square.

> From the song "Einstein A-go-go," by the pop group Landscape

It is as important an event as would be the transfer of the Vatican from Rome to the New World. The pope of physics has moved, and the United States will now become the center of the natural sciences.

Paul Langevin, on Einstein's move to America. Quoted in Pais, *A Tale of Two Continents*, 227

Anyone who has had the pleasure of being close to Einstein knows that he is not surpassed by anyone in respecting the intellectual property of others, in personal modesty, and in his distaste for publicity.

Max von Laue, Walther Nernst, and Heinrich Rubens, in a joint declaration of support for Einstein as anti-Semitism and anti-relativism were spreading among German physicists. *Tägliche Rundschau*, August 26, 1920

Virtually nothing but Einstein is being talked about here [in England], and if he were to come over now, I think he would be celebrated like a victorious general. The fact that a German's theory has been confirmed by observations made by the English has, as is becoming more obvious every day, brought the possibility of cooperation between these nations far closer. Thus Einstein, quite apart from the high scientific value of his inspired theory, has done an inestimable service to humankind.

Robert Lawson, who later translated the U.K. edition of *The Meaning of Relativity* (1922), in a letter to Arnold Berliner, who repeated it in a letter to Einstein on November 29, 1919; on the reception of the confirmation of general

relativity theory. Lawson was a lecturer in physics at the
University of Sheffield. Einstein Archives 7-004

*The foremost example of the damaging influence
upon natural science from the Jewish side was presented by Mr. Einstein, with his "theories" mathematically blundered together out of good, pre-existing
knowledge and his own arbitrary garnishes, theories which are now gradually decaying, which is the
fate of procreations alien to nature. In the process,
one cannot spare researchers, even those with genuine achievements, from the charge that they indeed
first let the "relativity-Jews" become established in
Germany. They did not see—nor did they want to
see—how very erroneous it was, even in a non-
scholarly connection, to consider especially this Jew
as a "good German."

> German physicist and winner of the 1905 Nobel Prize,
> Philipp Lenard, in *Völkischer Beobachter* 46 (May 13,
> 1933)

Jewish physics can best and most justly be characterized by recalling the activity of one who is probably its most prominent representative, the pure-
blooded Jew, Albert Einstein. His relativity theory
was supposed to transform all of physics, but when
faced with reality it did not have a leg to stand on. In
contrast to the intractable and solicitous desire for
truth in the Aryan scientist, the Jew lacks to a striking degree any comprehension of truth.

Philipp Lenard, in his book, *German Physics* (Munich: Lehmann's Verlag, 1936). Earlier in the century, Lenard and Einstein had had great respect for each other, then came into conflict over the general theory of relativity. Lenard's experiments on the photoelectric effect had led Einstein toward the hypothesis of light quanta.

With Albert Einstein died someone who vindicated the honor of mankind and whose name will never be forgotten.

Thomas Mann, "On the Death of Albert Einstein," in *Autobiographisches* (Frankfurt am Main: Fischer Bücherei, 1968)

It was interesting to see them together—Tagore, the poet with the head of a thinker, and Einstein, the thinker with the head of a poet. It seemed to an observer as though two planets were engaged in a chat.

Journalist Dmitri Marianoff, Margot Einstein's husband, to the *New York Times*, on his observations of a conversation between Einstein and Indian poet, musician, and mystic Rabindranath Tagore on July 14, 1930. See Tagore, "Farewell to the West" (1930–1931), 294–295; idem, *The Religion of Man* (New York: Macmillan, 1931), Appendix 2, 221–225

My husband is in Salzburg at present at a meeting of German scientists, where he will present a lecture. He now belongs to the circle of the most outstanding physicists of German origin. I am very happy about his success, which he really deserves.

Mileva Einstein-Marić to Helene Savić, September 3, 1909. In Roboz Einstein, *Hans-Albert Einstein*, 95

Albert is now a very famous physicist and respected in the world of physics. . . . Albert has devoted himself completely to physics and it seems to me that he has little time if any for the family.

> Ibid., March 12, 1913, 96

Anyone who advises Americans to keep secret information which they may have about spies and saboteurs is himself an enemy of America.

> Senator Joseph McCarthy, regarding Einstein's advocacy of refusing to testify at the House Un-American Activities Committee hearings. *New York Times*, June 14, 1953

Our travellers tell us that there is complete ignorance in the public mind as to what relativity means. A good many people seem to think that the book deals with the relations between the sexes.

> From Methuen publishers in England to Robert Lawson, Einstein's translator, February 1920. See *CPAE*, Vol. 9, Doc. 326

The mathematical education of the young physicist Albert Einstein was not very solid, which I am in a good position to evaluate since he obtained it from me in Zurich a long time ago.

> Hermann Minkowski, quoted on the Internet, www.gap .dcs.st-and.ac.uk/~history/Quotations/Minkowski.html (I usually don't rely on the Internet's undocumented quotations, but I hope this one is correct because I like it.)

He wore his usual jersey, baggy pants, and slippers. What especially struck me as he approached the doorway was that he seemed not to walk but to glide in a sort of undeliberate dance. It was enchanting. And there he was, bright, sad eyes, cascading white hair, with a smile of greeting on his face, a firm handshake.

From anthropologist Ashley Montagu's "Conversations with Einstein," *Science Digest*, July 1985

The great scientist of our age, he was truly a seeker after truth who would not compromise with evil or untruth.

Jawaharlal Nehru, prime minister of India, 1955

*I believe that so far as the development of physics is concerned, we can be very happy to have found so original a young thinker. . . . Einstein's "quantum hypothesis" is probably one of the most remarkable ever devised. . . . If it is false, well, then it will remain for all time "a beautiful memory."

German physical chemist and 1920 Nobel laureate Walther Nernst, in a March 17, 1910, letter to Arthur Schuster. *CPAE*, Vol. 3, xxiii, n. 36. Translated in Neffe, *Einstein*, 330

When he offered his last important work to the publishers, he warned them that there were no more than twelve persons in the whole world who would understand it, but the publishers took the risk.

This false report by a *New York Times* reporter, November 10, 1919, regarding the general theory of relativity, became established in the Einstein mythology. On December 3, 1919, another *New York Times* reporter asked if this statement was true, upon which "the doctor laughed good-humoredly." See Fölsing, *Albert Einstein*, 447, 451

It flashed to the minds of men the most spectacular proof of the Einstein Theory of Relativity, which provided the key to the vast treasure house of energy within the atom.

New York Times, August 7, 1945, on the atom bomb

This man changed thinking about the world as only Newton and Darwin changed it.

In the *New York Times* after Einstein's death, 1955. Quoted in Cahn, *Einstein*, 120

"For his unique services to theoretical physics and in particular for his discovery of the law of the photoelectric effect."

Nobel Prize Committee, official citation for the Nobel Prize in Physics, 1921. Note that no mention was made of relativity theory, which was still a controversial topic at the time. Einstein had been nominated for the prize each year from 1910 to 1918 except for 1911 and 1915. See Pais, *Subtle Is the Lord*, 505

He has a quiet way of walking, as if he is afraid of alarming the truth and frightening it away.

Japanese cartoonist Ippei Okamoto on Einstein's visit to
Japan, November 1922. See manuscript, "Einstein's 1922
Visit to Japan," in Einstein Archives 36-409

*Few men have contributed so much to our under-
standing of the physical world. . . . We see in Einstein,
especially those who have come to know him a little,
those personal qualities that are the counterpart of
great work: selflessness, humor, and deep kindness.

Robert Oppenheimer, March 16, 1939, in a radio address on
the occasion of Einstein's sixtieth birthday. Reprinted in
Science 89 (1939), 335–336

It is not too soon to start to dispel the clouds of myth
and to see the great mountain peak that these clouds
hide. As always, the myth has its charms, but the
truth is far more beautiful.

Robert Oppenheimer on Einstein, 1965. Quoted in the *Los
Angeles Times*, editorial, March 14, 1979, on the centennial
of Einstein's birth

He was almost wholly without sophistication and
wholly without worldliness. . . . There was always
in him a powerful purity at once childlike and profoundly stubborn.

Robert Oppenheimer, in "On Albert Einstein," *New York
Review of Books*, March 17, 1966

*He spent [his last 25 years] first trying to prove
that the quantum theory had inconsistencies in it.

No one could have been more ingenious in thinking up unexpected and clever examples; but it turned out the inconsistencies were not there; and often their resolution could be found in earlier work of Einstein himself. When that did not work . . . Einstein had simply to say that he did not like the theory.

Ibid.

*He fought with Bohr in a noble and furious way, and he fought with the theory which he had fathered but which he hated. It was not the first time that this had happened in science.

Ibid.

*Even above his humanity and kindliness, even above his immense analytical power and depth he had a quality that made him unique. This was his faith that there exists in the natural world an order and a harmony and that this may be apprehended by the mind of man.

Robert Oppenheimer in a eulogy for Einstein for the *Princeton Packet*, April 1955. Oppenheimer Papers, Box 256, Library of Congress. But according to Schweber, *Einstein and Oppenheimer*, 276, "Oppenheimer made much less complimentary statements: that Einstein had no understanding of or interest in modern physics . . . wasting his time in trying to unify gravitation and electromagnetism. . . . Oppenheimer complained that even though the Institute [of Advanced Study] had supported Einstein for twenty-five years," he left all his papers to the Hebrew University in Israel.

He responded with one of the most extraordinary kinds of laughter. . . . It was rather like the barking of a seal. It was a happy laughter. From that time on, I would save a good story for our next meeting, for the sheer pleasure of hearing Einstein's laugh.

> Abraham Pais. Quoted in Jeremy Bernstein, *Einstein* (Penguin, 1978), 77

*He was the freest man I have ever known. . . . Better than anyone before or after him, he knew how to invent invariance principles and make use of statistical fluctuations.

> Pais, *Subtle Is the Lord*, vii

What Einstein said wasn't all that stupid.

> Physicist and future Nobelist Wolfgang Pauli as a student, after hearing Einstein, twenty years his senior, give a lecture. Quoted in Ehlers, *Liebes Hertz!* 47

I'll never forget this speech. He was like a king who abdicated and installed me as his choice for successor.

> Wolfgang Pauli, recalling Einstein's speech at a dinner honoring Pauli after he won the Nobel Prize. Einstein said he is at the end of his wisdom, and it is now up to Pauli to further pursue a unified field theory. Quoted in Armin Hermann, "Einstein und die Österreicher," *Plus Lucis*, February 1995, 20–21

Doctor with the bushy head
Tell us that you're not a Red.

Tell us that you do not eat
Capitalists in the street.
Say to us it isn't true
You devour their children too.
Speak, oh speak, and say you're notsky
Just a bent-space type of Trotsky.

Verse written by popular newspaper columnist H. I. Phillips during the McCarthy era, poking fun at anti-Communist opposition to Einstein's admittance to the United States two decades earlier. Quoted by Norman F. Stanley in *Physics Today*, November 1995, 118

You have long dwelt in our midst as a very near and personal reminder of our highest aspirations; for this gentle and all-pervasive influence we are particularly grateful to you.

From a seventy-fifth birthday letter from the Physics Department at Princeton University, signed by Robert Dicke, Eugene Wigner, John Wheeler, Valentin Bargmann, Arthur Wightman, George Reynolds, Frank Shoemaker, Eric Rogers, Sam Treiman, and others, March 12, 1954. Einstein Archives 30-1242

In boldness it exceeds anything so far achieved in speculative natural science, in philosophical cognition theory. Non-Euclidean geometry is child's play by comparison.

Max Planck on Einstein's definition of time, in a lecture delivered at Columbia University, Spring 1909 (published Leipzig, 1910), 117ff.

Even though in political matters a deep gulf divides us, I am also absolutely certain that in the centuries to come Einstein will be celebrated as one of the brightest stars that ever shined on our academy.

> Max Planck to Heinrich von Ficker, March 31, 1933, on Einstein's resignation from the Prussian Academy of Sciences. Quoted in Christa Kirsten and H.-J. Treder, *Albert Einstein in Berlin, 1913–1933* (Berlin, 1979)

*It is difficult to make him your enemy, but once he has cast you out of his heart, you are done for as far as he is concerned.

> Einstein's doctor János Plesch, quoted in Herneck, *Einstein privat* (Berlin, 1976), 89

Einstein loved women, and the commoner and sweatier and smellier they were, the better he liked them.

> Peter Plesch, quoting his father, János. In Highfield and Carter, *The Private Lives of Albert Einstein*, 206

What we must particularly admire in him is the facility with which he adapts himself to new concepts and that he knows how to draw from them every conclusion.

> Henri Poincaré, 1911. Quoted in Hoffmann, *Albert Einstein: Creator and Rebel*, 99

*Einstein wanted a physical world that was unproblematically objective and deterministic, hence his rejection of modern quantum theory. This stance made him the last of the great ancients, rather than the first of the moderns.

Physicist and Anglican priest John Polkinghorne, quoted in *Science and Theology News* (online), November 18, 2005

*Nature and Nature's laws lay hid in night. God said, *Let Newton Be!* and all was light.

It did not last: the Devil howling "Ho! Let Einstein be!" restored the status quo.

God rolled his dice, to Einstein's great dismay: "Let Feynman Be!" and all was clear as day.

The first two lines were by Alexander Pope, in Epitaph XII for Sir Isaac Newton (see *The Works of Alexander Pope* [1797], Vol. 2, 403), to which Sir John Collings Squire, the literary editor of *The New Statesman* and *The London Mercury*, added the second couplet in his "Continuation of Pope on Newton," in *Poems* (1926). The third was added by historian of science Stephen Brush in his review, "Feynman's Success: Demystifying Quantum Mechanics," of Jagdesh Mehra's biography of Richard Feynman (*The Beat of a Different Drum*, 1994) in *American Scientist* 83 (September–October 1995), 477. (My thanks to Professor Brush for clarifying the provenance of the final couplet.)

A great transformer of natural science has been lost.

Pravda, Moscow, after Einstein's death. Quoted in Cahn, *Einstein*, 121

Instead of having our boys come home, this foreign-born agitator would have us plunge into another European war in order to further the spread of communism throughout the world. . . . It is about time the American people got wise to Einstein. In my opinion he is violating the law and ought to be prosecuted. . . . I call upon the Department of Justice to put a stop to this man Einstein.

> Congressman John Rankin (D-Miss.), *Congressional Record-House*, October 25, 1945. Facsimile reprinted in Cahn, *Einstein*, 101. Einstein had sent a letter to Congress asking for money for the American Committee for Spanish Freedom to carry on the fight for breaking relations with Franco's Spain, which Rankin and others considered a Communist plot.

I could hardly suggest that he spruce himself up for my camera, though on several occasions I did try surreptitiously to brush his hair back behind his ears—accidentally, as it were—while arranging the lights behind his hair. But the unruly swatch always sprang forward again with a stubbornness of its own. I gave up on his hair, but his feet continued to bother me. . . . Professor Einstein seldom wore socks. Although I tried to take all his pictures from the knees or waist up, it was difficult to keep my eyes off those bare ankles.

> Princeton University photographer Alan Richards, who had been called to take Einstein's official birthday portrait. See Richards, "Reminiscences," in *Einstein as I Knew Him*

Once, when a company had sent him a very sizable consulting fee, he used the check for a bookmark, then lost the book.

> Ibid.

He simplified his concerns in order to spend his time wisely. . . . This same uncluttered attitude allowed him to speak directly, with unaffected kindness and respect, to every human being he met, child or adult, ignoring externals.

Ibid.

*Einstein is more dangerous as a proponent of a cause than an enemy of that cause. His genius is limited to science. In other matters he is a fool. . . . He should just stay out of it! He is made only for his equations.

Pacifist Romain Rolland in a letter to writer Stefan Zweig, September 15, 1933. Quoted in Grüning, *Ein Haus für Albert Einstein*, 386–387

Einstein, a genius in science, is weak, indecisive, and contradictory outside his own field. . . . His continuous change of opinion and . . . change in his actions are worse than the inflexible obstinacy of a declared enemy.

Pacifist Romain Rolland, diary entry of September 1933. Quoted in Nathan and Norden, *Einstein on Peace*, 233

*Sometimes you felt Einstein was creating his ideas while he spoke, forgetting that . . . he had an audience. Then, when he realized we were present, he gave one of his striking and drastic, always appropriate similes.

Ilse Rosenthal-Schneider, his former student, in *Reality and Scientific Truth*, 91

> To Einstein, hair and violin,
> We give our final nod.
> He's understood by just two folks:
> Himself . . . and sometimes God.

An ode by Jack Rossetter, sent by a reader from India

In an age when physics has produced a large number of great men and a bewildering variety of new facts and theories, Einstein remains supreme in the breadth and depth and comprehensiveness of his constructions.

Bertrand Russell, ca. 1928, in an unpublished article in the Russell Archive, McMaster University, Hamilton, Ontario, Canada. Einstein Archives 33-154

Einstein was indisputably one of the greatest men of our time. He had, in a high degree, the simplicity characteristic of the best men in science—a simplicity which comes of a single-minded desire to know and understand things that are completely impersonal.

Bertrand Russell. In *The New Leader*, May 30, 1955

He removed the mystery from gravitation, which everybody since Newton had accepted, with a reluctant feeling, as unintelligible.

Bertrand Russell. Quoted in Whitrow, *Einstein*, 22

Of all the public figures that I have known, Einstein was the one who commanded my most whole-hearted admiration. . . . Einstein was not only a great scientist, he was a great man. He stood for peace in a world drifting towards war. He remained sane in a mad world, and liberal in a world of fanatics.

> Ibid., 90

One student, by name of Einstein, even sparkled in rendering an adagio from a Beethoven sonata with deep understanding.

> Music inspector J. Ryffel of the Aargau cantonal school in evaluating Einstein's performance at a final music exam at the school, ca. March 31, 1896. *CPAE*, Vol. 1, Doc. 17

For heaven's sake, Albert, can't you *count*?

> Pianist Artur Schnabel, after Einstein made several wrong entrances in a quartet rehearsal at the Mendelssohn villa in Berlin in the 1920s. Recalled by Mike Lipskin and quoted by Herb Caen in the *San Francisco Chronicle*, February 3, 1996

Even though without writing each other, we are in mental communication; for we respond to our dreadful times in the same way and tremble together for the future of mankind. . . . I like it that we have the same given name.

> Albert Schweitzer. In a letter to Einstein, February 20, 1955. Einstein Archives 33-236

Einstein

A little mousey man he was
With board, and chalk in hand;
And millions were awestruck because
They couldn't understand.
Said he: "E equals mc 2:
I'll prove it true.
No doubt you can, you marvel man,
But will it serve our good?
Will it prolong our living span
And multiply our food?
Will it bring peace between the nations
To make equations?
Our thanks are due no doubt to you
For truth beyond our ken;
But after all what did you do
To ease the lot of men?
How can a thousand "yous" be priced
Beside a Christ?

Poet Robert Service, who may not have been aware of
Einstein's humanitarian deeds. From *Later Collected Verse*
(New York: Dodd, Mead, 1965). (Used by permission of
the Estate of Robert Service, given by M. William
Krasilovsky)

A powerful searchlight of the human mind, piercing
by its rays the darkness of the unknown, has suddenly been extinguished. The world has lost its

foremost genius and the Jewish people its most illustrious son in the present generation.

> Moishe Sharrett, prime minister of Israel, 1955. Quoted in
> Cahn, *Einstein*, 120

You are the only sort of man in whose existence I see
much hope for in this deplorable world.

> George Bernard Shaw, in a postcard of December 2, 1924.
> Einstein Archives 33-242

Tell Einstein that I said the most convincing proof I
can adduce of my admiration for him is that his is the
only one of these portraits [of celebrities] I paid for.

> George Bernard Shaw. Recalled by Archibald Henderson in
> the *Durham Morning Herald*, August 21, 1955. Einstein Ar-
> chives 33-257. Einstein's reply: "That is very characteristic
> of Bernard Shaw, who has declared that money is the most
> important thing in the world."

Ptolemy made a universe, which lasted 1400 years.
Newton, also, made a universe, which lasted 300
years. Einstein has made a universe, and I can't tell
you how long that will last.

> George Bernard Shaw, at a banquet in England honoring
> Einstein. Quoted in David Cassidy, *Einstein and Our World*
> (Humanities Press, 1995), 1. Also shown in *Nova*'s Einstein
> biography for PBS television, 1979

There is only one fault with his cosmical religion: he
put an extra letter in the word—the letter "s."

Bishop Fulton J. Sheen. Quoted in Clark, *Einstein*, 426. Even though Sheen seems to have disapproved of his religious views, Einstein still admired him. See Fantova, "Conversations with Einstein," December 13, 1953, where he praises Sheen.

What did surprise me was his physique. He had come in from sailing and was wearing nothing but a pair of shorts. It was a massive body, very heavily muscled; he was running to fat round the midriff and in the upper arms, rather like a footballer in middle age, but he was an unusually strong man.

C. P. Snow on his visit to Einstein in 1937. Quoted by Richard Rhodes, *The Making of the Atom Bomb* (New York: Touchstone Books, 1995)

To me, he appears as out of comparison the greatest intellect of this century, and almost certainly the greatest personification of moral experience. He was in many ways different from the rest of the species.

C. P. Snow, in "Conversations with Einstein," quoted in French, *Einstein: A Centenary Volume*, 193

I loved him and admired him profoundly for his basic goodness, his intellectual genius and his indomitable moral courage. In contrast to the lamentable vacillation that characterizes most so-called intellectuals, he fought tirelessly against injustice and evil. He will live in the memory of future generations not only as a scientific genius of exceptional stature but also as an epitome of moral greatness.

> Lifelong friend Maurice Solovine in his introduction to *Letters to Solovine*. Quoted in Abraham Pais, "Albert Einstein as Philosopher and Natural Scientist," 1956

*Tall, robust, of an almost Balzacian physique, but with a pale face of the most pure and sweet Oriental pallor and with two deep, thoughtful and melancholy eyes whose pupils seem to reflect all of the spiritual defeats and ascents of past Jewish generations that survived the exhausting tumult of martyrdom and anxiety, Albert Einstein . . . answered my questions with tired and kindly attention.

> Aldo Sorani, in the Italian newspaper *Il Messagero*, October 26, 1921. *CPAE*, Vol. 12, Appendix G

*It was like going to tea with god, not the terrible God of the Bible, but the little child's father in heaven, very kind and wise. Yet Einstein himself was very much like a child.

> Liberal journalist I. F. Stone. Quoted in Brian, *Einstein*, 403

He was a Zionist on general humanitarian grounds rather than on nationalistic grounds. He felt that Zionism was the only way in which the Jewish problem in Europe could be settled. . . . He was never in favor of aggressive nationalism, but he felt that a Jewish homeland in Palestine was essential to save the remaining Jews in Europe. . . . After the State of Israel was established, he said that somehow he felt happy he was not there to be involved in the deviations from the high moral tone he detected.

Ernst Straus. Quoted in Whitrow, *Einstein*, 87–88

I would be unable to picture science without him. His spirit permeates it. He makes part of my thinking and outlook.

Albert Szent-Gyorgy, Nobel laureate in medicine and physiology. Quoted in Cahn, *Einstein*, 122

His shock of white hair, his burning eyes, his warm manner again impressed me with the human character of this man who dealt so abstractly with the laws of geometry and mathematics. . . . There was nothing stiff about him—there was no intellectual aloofness. He seemed to me a man who valued human relationships and he showed toward me a real interest and understanding.

Indian poet, musician, artist, and mystic Rabindranath Tagore, after his meetings with Einstein in Germany in 1930. Quoted in the *New York Times*, August 20, 2001

Nobody in *football* should be called a genius. A genius is a guy like Norman Einstein.

Football commentator and former player Joe Theisman. Quoted in *The Book of Truly Stupid Sports Quotes* (New York: HarperCollins, 1996)

One of the greatest—perhaps *the* greatest—of achievements in the history of human thought.

Joseph John Thomson, discoverer of the electron, referring to Einstein's work on general relativity, 1919. Quoted in Hoffmann, *Albert Einstein: Creator and Rebel*, 132

As the century's greatest thinker, as an immigrant who fled from oppression to freedom, as a political idealist, he best embodies what historians will regard as significant about the twentieth century. And as a philosopher with faith both in science and in the beauty of God's handiwork, he personifies the legacy that has been bequeathed to the next century.

Time magazine, explaining its selection of Einstein as Person of the Twentieth Century, January 3, 2000

He had the kind of male beauty that, especially at the beginning of the century, caused great commotion.

Antonina Vallentin, in *Le Drame d'Albert Einstein* (Paris, 1954) and *Das Drama Albert Einsteins* (Stuttgart, 1955), 9

*You had only to see Einstein in a small sailboat to realize the strength of the roots that bound him to a simple open-air life. Wearing sandals and an old sweater, his hair ruffling in the breeze, he would . . . rock gently with the motion of the boat, completely at one with the sail he was maneuvering. . . . As he tugged at the sail, with his muscles protruding like cables, . . . he might have belonged to the age of sea gods or pirates. . . . He looked like anything in the world but a scientist.

> Antonina Vallentin, in *The Drama of Albert Einstein* (Double-day, 1954), 168

*His shoulders were still robust, his bare neck strong and round. But time had dug deep furrows on the plump cheeks, and the lips drooped at the corners. . . . His high forehead was deeply wrinkled. . . . His hair was as wiry as ever, with that curious life of its own. . . . The most moving change, however, was in his eyes. The burning glance seemed to have singed the skin underneath. . . . But his strength burst forth through the eyes and triumphed over everything that was in decline.

> Ibid., 295, on seeing Einstein many years later, before his seventieth birthday

[Einstein] acted on women as a magnet acts on iron filings. But he also enjoyed the company of women and was captivated by everything feminine.

> Konrad Wachsmann, the architect of Einstein's house in Caputh. Quoted in Grüning, *Ein Haus für Albert Einstein*, 158

During our crossing, Einstein explained his theory to me every day, and by the time we arrived I was fully convinced he understood it.

> Chaim Weizmann, Spring 1921, after he escorted Einstein to the United States on the SS *Rotterdam* on behalf of a

Zionist delegation. Quoted in Seelig, *Albert Einstein and die Schweiz*, 82

[Einstein] is acquiring the psychology of a prima donna who is beginning to lose her voice.

Chaim Weizmann, 1933, in response to Einstein's requests for reforms at Hebrew University. Quoted in Norman Rose, *Chaim Weizmann* (New York, 1986), 297

The world has lost an illustrious scientist, a great and brave mind and a fighter for human rights. The Jewish people have lost the brightest jewel in their crown.

Vera Weizmann, widow of the late Israeli president, after Einstein's death, 1955. Quoted in Cahn, *Einstein*, 121

Einstein was a physicist and not a philosopher. But the naïve directness of his questions was philosophical.

C. F. von Weizsaecker. Quoted in P. Aichelburg and R. Sexl, *Albert Einstein* (Vieweg, 1979), 159

So today, for his genius and integrity we, who inadequately measure his power, salute the new Columbus of science voyaging through the strange seas of thought alone.

Dean Andrew Fleming West of Princeton University, after reading a citation before President John Grier Hibben conferred an honorary doctorate on Einstein, May 9, 1921

Princeton Alumni Weekly, May 11, 1921, 713–714. See also Illy, *Albert Meets America*, 166

*Of all the questions with which the great thinkers have occupied themselves in all lands and all centuries, none has ever claimed greater primacy than the origin of the universe, and no contributions to this issue ever made by any man anytime have proved themselves richer in illuminating power than those that Einstein made.

Princeton physicist John Wheeler in "Einstein," *Biographical Memoirs of the National Academy of Sciences* 51 (1980), 97

The Einstein and the Eddington

The sun was setting on the links,
The moon looked down serene,
The caddies all had gone to bed,
But still there could be seen
Two players lingering by the trap
That guards the thirteenth green.
The Einstein and the Eddington
Were counting up their score;
The Einstein's card showed ninety-eight
And Eddington's was more.
And both lay bunkered in the trap
And both stood there and swore.
I hate to see, the Einstein said;
Such quantities of sand;
Just why they placed a bunker here

I cannot understand.
If one could smooth this landscape out,
I think it would be grand.
If seven maids with seven mops
Would sweep the fairway clean
I'm sure that I could make this hole
In less than seventeen.
I doubt it, said the Eddington,
Your slice is pretty mean.
Then all the little golf balls came
To see what they were at,
And some of them were tall and thin
And some were short and fat,
A few of them were round and smooth,
But most of them were flat.
The time has come, said Eddington,
To talk of many things:
Of cubes and clocks and meter-sticks
And why a pendulum swings.
And how far space is out of plumb,
And whether time has wings.
I learned at school the apple's fall
To gravity was due,
But now you tell me that the cause
Is merely G-mu-nu,
I cannot bring myself to think
That this is really true.
You say that gravitation's force
Is clearly not a pull.
That space is mostly emptiness,

While time is nearly full;
And though I hate to doubt your word,
It sounds like a bit of bull.
And space, it has dimensions four,
Instead of only three.
The square of the hypotenuse
Ain't what it used to be.
It grieves me sore, the things you've done
To plane geometry.
You hold that time is badly warped,
That even light is bent:
I think I get the idea there,
If this is what you meant:
The mail the postman brings today,
Tomorrow will be sent.
If I should go to Timbuctoo
With twice the speed of light,
And leave this afternoon at four,
I'd get back home last night.
You've got it now, the Einstein said,
That is precisely right.
But if the planet Mercury
In going round the sun,
Never returns to where it was
Until its course is run,
The things we started out to do
Were better not begun.
And if before the past is through,
The future intervenes;
Then what's the use of anything;

Of cabbages or queens?
Pray tell me what's the bally use
Of Presidents and Deans.
The shortest line, Einstein replied,
Is not the one that's straight;
It curves around upon itself,
Much like a figure eight,
And if you go too rapidly
You will arrive too late.
But Easter day is Christmas time
And far away is near,
And two and two is more than four
And over there is here.
You may be right, said Eddington,
It seems a trifle queer.
But thank you very, very much,
For troubling to explain;
I hope you will forgive my tears,
My head begins to pain;
I feel the symptoms coming on
Of softening of the brain.

Dr. W. H. Williams, who shared an office with Arthur Eddington at the University of California at Berkeley, prepared this verse for a faculty club dinner on the eve of Eddington's departure from Berkeley in 1924. This poem is, of course, based on "The Walrus and the Carpenter" in Lewis Carroll's *Through the Looking-Glass and What Alice Found There* (1872).

He is not a good teacher for mentally lazy gentlemen who merely want to fill up a notebook and then learn it by heart for an exam; he is not a smooth talker. But anyone who wants to learn how to construct physical ideas, carefully examine all premises, take note of the pitfalls and problems, review the reliability of his reflections, will find Einstein a first-rate teacher.

> Heinrich Zangger, in a letter to Ludwig Forrer, October 9, 1911, recommending Einstein for a post at the ETH in Zurich. *CPAE*, Vol. 5, Doc. 291

An Einstein would long since have been hanged as a mystic because for him light deflects around corners.

> Heinrich Zangger, October 17, 1919, on conditions in Russia and referring to the recent proof of general relativity

Einstein's [violin] playing is excellent, but he does not deserve his world fame; there are many others just as good.

> A music critic on an early 1920s performance, unaware that Einstein's fame derived from physics, not music. Quoted in Reiser, *Albert Einstein*, 202–203

"Prof. Einstein's Got a New Baby: Formula Keeps Our Man Up Nights."

> Headline of a book review of *The Meaning of Relativity* which appeared in the *Daily Mirror* (New York), March 30, 1953. It referred to the appendix published two years before Einstein's death in which the physicist presented a

greatly simplified derivation of the equations of general relativity. (Contributed by Trevor Lipscombe)

Weird Math

Said Einstein, "I have an equation,
Which some might call Rabelaisian:
Let P be virginity,
Approaching infinity,
And let U be a constant, persuasion
"Now, if P over U be inverted,
And the square root of U be inserted
X times over P,
The result, Q.E.D.,
Is a relative," Einstein asserted

By an anonymous wag, found on the Internet, November 11, 2003

Teutonic Einstein

Here lies Einstein, an enterprising Teuton
Who, relatively speaking, silenced Newton.

Epitaph for Einstein by an unidentified author. Quoted by Ashley Montagu in "Conversations with Einstein," *Science Digest*, July 1985. Einstein might have taken issue with the "Teuton" characterization.

The Steins (two versions)

Three wonderful people called Stein;
There's Gert and there's Ep and there's Ein.
Gert writes in blank verse,

Ep's sculptures are worse,
And nobody understands Ein.

Verse by an unidentified author. Quoted in ibid.

I don't like the family Stein!
There is Gert, there is Ep, there is Ein.
Gert's writings are punk,
Ep's statues are junk,
Nor can anyone understand Ein.

Rhyme current in the United States in the 1920s. Listed
in the *Oxford Dictionary of Quotations* (1999) under
"Anonymous"

Fast Girl

There was a young lady named Bright
Whose speed was much faster than light.
She went out one day
In a relative way
And came back the previous night.

Limerick about relativity, ca. 1919, unknown author, in *New
Statesman*, August 9, 1999 (found on its Web site)

And, finally:

All boys are idiots except for Albert Einstein.

Mary Lipscombe, age 8, to Lottie Appel, age 6, daughters
of my former colleagues Trevor and Fred, respectively.
Overheard at Princeton University Press Christmas party,
December 21, 1999

Bibliography

Abraham, Carolyn. *Possessing Genius: The Bizarre Odyssey of Einstein's Brain*. New York:

Born, Max, ed. *Einstein-Born Briefwechsel, 1916–1955*. Munich: Nymphenbürger, 1969.

———. *The Born-Einstein Letters*. Trans. Irene Born. New York: Macmillan, 2005.

Brian, Denis. *Einstein, a Life*. New York: Wiley, 1996.

Brockman, John. *My Einstein: Essays by Twenty-four of the World's Leading Thinkers on the Man, His Work, and His Legacy*. New York: Pantheon, 2006.

Buchwald, Diana Kormos, Ze'ev Rosenkranz, et al., eds., *The Collected Papers of Albert Einstein*, Vol. 12, *The Berlin Years: Correspondence, January–December 1921*. Princeton, N.J.: Princeton University Press, 2009. (Trans. Ann Hentschel, 2009.)

Buchwald, Diana Kormos, Tilman Sauer, et al., eds., *The Collected Papers of Albert Einstein*, Vol. 10, *The Berlin Years: Correspondence, May–December 1920*. Princeton, N.J.: Princeton University Press, 2006. (Trans. Ann Hentschel, 2006.)

Buchwald, Diana Kormos, Robert Schulmann, et al., eds. *The Collected Papers of Albert Einstein*, Vol. 9, *The Berlin Years: Correspondence, January 1919–April 1920*. Princeton, N.J.: Princeton University Press, 2004. (Trans. Ann Hentschel, 2004.)

Cahn, William. *Einstein: A Pictorial Biography*. New York: Citadel Press, 1960.

Calaprice, Alice. *The Quotable Einstein*. Princeton, N.J.: Princeton University Press, 1996.

———. *The Expanded Quotable Einstein*. Princeton, N.J.: Princeton University Press, 2000.

———. *Dear Professor Einstein: Albert Einstein's Letters to and from Children*. Foreword by Evelyn Einstein. Amherst, N.Y.: Prometheus, 2002.

———. *The Einstein Almanac*. Baltimore: Johns Hopkins University Press, 2005.

Calaprice, Alice. *The New Quotable Einstein.* Princeton, N.J.: Princeton University Press, 2005.

Calaprice, Alice, and Trevor Lipscombe. *Albert Einstein: A Biography.* Greenwood Biographies (For young adults). Westport, Conn., and London: Greenwood Press, 2005.

Cassidy, David. *Einstein and Our World.* Atlantic Highlands, N.J.: Humanities Press, 1995.

Clark, Ronald W. *Einstein: The Life and Times.* New York: World Publishing, 1971.

CPAE. See Stachel et al. for Vols. 1 and 2; Klein et al. for Vol. 5; Kox et al. for Vol. 6; Janssen et al. for Vol. 7; Schulmann et al. for Vol. 8; Buchwald et al. for Vols. 9, 10, 12.

Dukas, Helen, and Banesh Hoffmann. *Albert Einstein, the Human Side.* Princeton, N.J.: Princeton University Press, 1979.

Dürrenmatt, Friedrich. *Albert Einstein: Ein Vortrag.* Zurich: Diogenes, 1979.

Dyson, Freeman. "Writing a Foreword for Alice Calaprice's New Einstein Book." *Princeton University Library Chronicle* 57, no. 3 (Spring 1996), 491–502.

Ehlers, Anita. *Liebes Hertz!* Berlin: Birkhäuser, 1994.

Einstein, Albert. *Cosmic Religion with Other Opinions and Aphorisms.* New York: Covici-Friede, 1931.

———. *About Zionism.* Trans. L. Simon. New York: Macmillan, 1931.

———. *The World as I See It.* Abridged ed. New York: Philosophical Library, distributed by Citadel Press. Orig. in Leach, *Living Philosophies*, 1931.

———. *The Origins of the Theory of Relativity.* Glasgow: Jackson, Wylie, 1933.

———. *Essays in Science.* Trans. Alan Harris. New York: Philosophical Library, 1934.

———. *Mein Weltbild.* Amsterdam: Querido Verlag, 1934. Paperback ed., Berlin: Ullstein, 1993.

———. "Autobiographical Notes." In Schilpp, *Albert Einstein: Philosopher-Scientist.*

————. *Out of My Later Years*. Paperback ed. New York: Wisdom Library of the Philosophical Library, 1950. (Other editions exist as well; page numbers refer to this edition.)

————. *The Meaning of Relativity*. 5th ed. Princeton, N.J.: Princeton University Press, 1953. Includes Appendix to 2d ed.

————. *Ideas and Opinions*. Trans. Sonja Bargmann. New York: Crown, 1954. (Other editions exist as well; page numbers refer to this edition.)

————. *Correspondance avec Michèle Besso, 1903–1955*. Paris: Hermann, 1979.

————. *Albert Einstein/Mileva Marić: The Love Letters*. Ed. Jürgen Renn and Robert Schulmann. Trans. Shawn Smith. Princeton, N.J.: Princeton University Press, 1992.

————. *Einstein on Humanism*. New York: Carol Publishing, 1993.

————. *Letters to Solovine, 1906–1955*. Trans. from the French by Wade Baskin, with facsimile letters in German. New York: Carol Publishing, 1993.

————. *Relativity: The Special and the General Theory*. London: Routledge Classics, 2001.

Einstein, Albert, and Sigmund Freud. *Why War?* Paris: Institute for Intellectual Cooperation, League of Nations, 1933.

Einstein, Albert, and Leopold Infeld. *The Evolution of Physics*. New York: Simon & Schuster, 1938.

Einstein: A Portrait. Introduction by Mark Winokur. Corte Madera, Calif.: Pomegranate Artbooks, 1984.

Fantova, Johanna. "Conversations with Einstein," October 1953–April 1955. Manuscript. Fantova Collection on Albert Einstein. Manuscripts Division. Department of Rare Books and Special Collections. Princeton University Library, Princeton, N.J.

Flückiger, Max. *Albert Einstein in Bern*. Bern: Haupt, 1972.

Fölsing, Albrecht. *Albert Einstein*. Trans. Ewald Osers. New York: Viking, 1997.

Frank, Philipp. *Einstein: His Life and Times*. New York: Knopf, 1947, 1953.

——. *Einstein: Sein Leben und seine Zeit*. Braunschweig, Germany: Vieweg, 1979.

French, A. P., ed. *Einstein: A Centenary Volume*. Cambridge, Mass.: Harvard University Press, 1979.

Galison, Peter L., Gerald Holton, and Silvan S. Schweber. *Einstein for the Twenty-first Century*. Princeton, N.J.: Princeton University Press, 2008.

Grüning, Michael. *Ein Haus für Albert Einstein*. Berlin: Verlag der Nation, 1990.

Hadamard, Jacques. *An Essay on the Psychology of Invention in the Mathematical Field*. Princeton, N.J.: Princeton University Press, 1945.

Highfield, Roger, and Paul Carter. *The Private Lives of Albert Einstein*. London: Faber and Faber, 1993.

Hoffmann, Banesh. *Albert Einstein: Creator and Rebel*. New York: Viking, 1972.

——. "Einstein and Zionism." In *General Relativity and Gravitation*, ed. G. Shaviv and J. Rosen. New York: Wiley, 1975.

Holton, Gerald. *The Advancement of Science and Its Burdens*. New York: Cambridge University Press, 1986.

Holton, Gerald, and Yehuda Elkana, eds. *Albert Einstein: Historical and Cultural Perspectives. The Centennial Symposium in Jerusalem*. Princeton, N.J.: Princeton University Press, 1982.

Hu, Danian. *China and Albert Einstein*. Cambridge, Mass.: Harvard University Press, 2005.

Illy, József. *Albert Meets America: How Journalists Treated Genius during Einstein's 1921 Travels*. Baltimore: Johns Hopkins University Press, 2006.

Infeld, Leopold. *The Quest: The Evolution of a Scientist*. New York: Doubleday, 1941.

——. *Albert Einstein*. Rev. ed. New York: Charles Scribner's Sons, 1950.

Isaacson, Walter. *Einstein: His Life and Universe*. New York: Simon & Schuster, 2007.

Jammer, Max. *Einstein and Religion*. Princeton, N.J.: Princeton University Press, 1999.

Janssen, Michel, Robert Schulmann, et al., eds. *The Collected Papers of Albert Einstein*, Vol. 7, *The Berlin Years: Writings, 1918–1921*. Princeton, N.J.: Princeton University Press, 2002. (Trans. Alfred Engel, 2002.)

Jerome, Fred. *The Einstein File: J. Edgar Hoover's Secret War against the World's Most Famous Scientist*. New York: St. Martin's, 2002.

———. *Einstein on Israel and Zionism*. New York: St. Martin's Press, 2009.

Jerome, Fred, and Rodger Taylor. *Einstein on Race and Racism*. New Brunswick, N.J.: Rutgers University Press, 2005.

Kaller's autographs catalog. "Jewish Visionaries," 1997. Kaller's Antiques and Autographs, at Macy's, 37th St., New York City.

Kantha, Sachi Sri. *An Einstein Dictionary*. Westport, Conn.: Greenwood Press, 1996.

———. "Medical Profile of Einstein: Six Analytical Papers." Available through author, Gifu University, Japan.

Klein, Martin, A. J. Kox, and Robert Schulmann, eds. *The Collected Papers of AlbertEinstein*, Vol. 5, *The Swiss Years: Correspondence, 1902–1914*. Princeton, N.J.: Princeton University Press, 1993. (Trans. Anna Beck, 1995.)

Kox, A. J., Martin J. Klein, and Robert Schulmann, eds. *The Collected Papers of Albert Einstein*, Vol. 6, *The Berlin Years: Writings, 1914–1917*. Princeton, N.J.: Princeton University Press, 1996. (Trans. Alfred Engel, 1996.)

Leach, Henry G., ed. *Living Philosophies: A Series of Intimate Credos*. New York: Simon & Schuster, 1931.

Levenson, Thomas. *Einstein in Berlin*. New York: Bantam, 2003.

Marianoff, Dmitri, with Palma Wayne. *Einstein: An Intimate Study of a Great Man*. Garden City, N.Y.: Doubleday, Doran, 1944.

Michelmore, P. *Einstein: Profile of the Man*. New York: Dodd, 1962.

Moszkowski, Alexander. *Conversations with Einstein*. Trans. Henry L. Brose. New York: Horizon Press, 1970. (Conversations took place in 1920, trans. 1921, published in English in 1970.)

Nathan, Otto, and Heinz Norden, eds. *Einstein on Peace*. New York: Simon & Schuster, 1960.

Neffe, Jürgen. *Einstein*. Trans. Shelley Frisch. New York: Farrar, Straus and Giroux, 2005.

Ohanian, Hans C. *Einstein's Mistakes: The Human Failings of Genius*. New York: Norton, 2008.

Pais, Abraham. *Subtle Is the Lord: The Science and the Life of Albert Einstein*. Oxford and New York: Oxford University Press, 1982.

———. *Einstein Lived Here*. Oxford and New York: Oxford University Press, 1994.

———. *A Tale of Two Continents*. Princeton, N.J.: Princeton University Press, 1997.

Planck, Max. *Where Is Science Going?* New York: Norton, 1932.

Popović, Milan, ed. *In Albert's Shadow: The Life and Letters of Mileva Marić, Einstein's First Wife*. Baltimore: Johns Hopkins University Press, 2003.

Regis, Ed. *Who Got Einstein's Office?* Reading, Mass.: Addison-Wesley, 1987.

Reiser, Anton. *Albert Einstein: A Biographical Portrait*. New York: Boni, 1930.

Richards, Alan Windsor. *Einstein as I Knew Him*. Princeton, N.J.: Harvest Press, 1979.

Robinson, Andrew. *Einstein: A Hundred Years of Relativity*. Bath, U.K.: Palazzo, 2005.

Roboz Einstein, Elizabeth. *Hans Albert Einstein: Reminiscences of His Life and Our Life Together*. Iowa City: Iowa Institute of Hydraulic Research, University of Iowa, 1991.

Rosenkranz, Ze'ev. *The Einstein Scrapbook*. Baltimore: Johns Hopkins University Press, 2002.

Rosenkranz, Ze'ev, and Barbara Wolff, eds. *Einstein: The Persistent Illusion of Transience*. Jerusalem: Magnes Press of Hebrew University, 2007.

Rosenthal-Schneider, Ilse. *Reality and Scientific Truth*. Detroit, Mich.: Wayne State University Press, 1980.

Rowe, David E., and Robert Schulmann. *Einstein on Politics: His Private Thoughts and Public Stands on Nationalism, Zionism, War, Peace, and the Bomb*. Princeton, N.J.: Princeton University Press, 2007.

Ryan, Dennis P., ed. *Einstein and the Humanities*. New York: Greenwood Press, 1987.

Sayen, Jamie. *Einstein in America*. New York: Crown, 1985.

Schilpp, Paul, ed. *Albert Einstein: Philosopher-Scientist*. Evanston, Ill.: Library of Living Philosophers, 1949.

Schulmann, Robert. "Einstein Rediscovers Judaism." Unpublished manuscript, 1999.

Schulmann, Robert, A. J. Kox., Michel Janssen, and József Illy, eds. *The Collected Papers of Albert Einstein*, Vol. 8, Parts A and B, *The Berlin Years: Correspondence, 1914–1918*. Princeton, N.J.: Princeton University Press, 1998. (Trans. Ann Hentschel, 1998.)

Schweber, Silvan S. *Einstein and Oppenheimer: The Meaning of Genius*. Cambridge, Mass.: Harvard University Press, 2008.

Seelig, Carl. *Albert Einstein und die Schweiz*. Zurich: Europa-Verlag, 1952.

Seelig, Carl, ed. *Helle Zeit, dunkle Zeit: In Memorium Albert Einstein*. Zurich: Europa Verlag, 1956.

Sotheby's auction catalog, June 26, 1998.

Stachel, John. "Einstein's Jewish Identity." Unpublished manuscript, 1989.

Stachel, John, ed., with the assistance of Trevor Lipscombe, Alice Calaprice, and Sam Elworthy. *Einstein's Miraculous Year: Five Papers That Changed the Face of Physics*. Princeton, N.J.: Princeton University Press, 1998.

Stachel, John, et al., eds. *The Collected Papers of Albert Einstein*, Vol. 1, *The Early Years: 1879–1902*. Princeton, N.J.:

Princeton University Press, 1987. (Trans. Anna Beck, 1987.)

Stachel, John, et al., eds. *The Collected Papers of Albert Einstein*, Vol. 2, *The Swiss Years: Writings, 1900–1909*. Princeton, N.J.: Princeton University Press, 1989. (Trans. Anna Beck, 1989.)

Stern, Fritz. *Einstein's German World*. Princeton, N.J.: Princeton University Press, 1999.

Viereck, George S. *Glimpses of the Great*. New York: Macauley, 1930.

Whitrow, G. J. *Einstein: The Man and His Achievement*. New York: Dover, 1967.

Index of Key Words

Subject Index

Page numbers in italics refer to illustrations.